PLANNING
TELECOMMUNICATION
NETWORKS

Books of Related Interest from IEEE Press...

PLANNING
TELECOMMUNICATION
NETWORKS

Thomas G. Robertazzi
State University of New York at Stony Brook

IEEE Communications Society, Sponsor

**IEEE
PRESS**

The Institute of Electrical and Electronics Engineers, Inc., New York

This book and other books may be purchased at a discount
from the publisher when ordered in bulk quantities. Contact:

IEEE Press Marketing
Attn: Special Sales
Piscataway, NJ 08855-1331
Fax: (732) 981-9334

For more information about IEEE PRESS products,
visit the IEEE Home Page: http//www.ieee.org/

Printed in the United States of America

10   9   8   7   6   5   4   3   2   1

ISBN 0-7803-4702-1
IEEE Order Number: PC5755

Library of Congress Cataloging-in-Publication Data

Robertazzi, Thomas G.
    Planning telecommunication networks / Thomas G. Robertazzi : IEEE
Communications Society, sponsor.
        p.   cm.
    Includes bibliographical references and index.
    ISBN 0-7803-4702-1
    1. Telecommunication systems – Planning.   I. IEEE Communications
Society.   II. Title.
TK5101.R55   1998   98–38908
621.382′1 – dc21   CIP

To Marsha, Rachel, and Deanna

CONTENTS

PREFACE

During the past two decades there has been very rapid growth in the number, sophistication, and capacity of networks used to tie together people and computers. This includes both telecommunications and computer networks. Investments in such equipment have grown in size, along with the serious consequences of network failure and capacity shortfall. Accordingly the need for sound network planning has also grown.

Network planning involves the orderly and efficient deployment and management of communication facilities *over time*. This temporal aspect is what distinguishes network planning from generic network management. However, it is sometimes useful to view the field of network planning as a distinct subset of the field of network management. Network planning draws on people and ideas from several fields, including electrical engineering, operations research, computer science, and applied mathematics.

Network planning was originally a child of the telephone industry. Most large telephone carriers have had network planning departments for some time. Telephone companies have been joined in their need for a planning function by cable TV operators and computer network operators. There has also been a steady introduction of new technologies requiring new planning. This includes computers in general, ISDN networks, SONET networks, cellular radio networks, WDM networks and fiber optics, ATM networks, new types of local area and metropolitan area networks, and new cable TV technology.

This book is an introduction to the theory of network planning at the senior/graduate level. It is intended for students, practitioners, and managers interested in the telephone, computer network, and cable TV industries. The book is written in the style of a textbook, complete with homework problems, and a solutions manual. This material is suitable as a text for either a one-quarter course on network planning or for half of a one-semester course on network management and planning. It is in this latter form that I have successfully taught this material to Stony Brook students since 1994.

In this work I have surveyed a number of fundamental areas that are crucial to modern network planning. These areas are mathematical programming, network algorithms, reliability, software and optimization, and data analysis. The chapters on the first three topics are largely quantitative. A good part, though not all, of the last two chapters is qualitative. This is particularly true of the sections dealing with new software technologies.

The wide variance in the type of material covered in this book is due to the multidisciplinary nature of network planning. For topics that are amenable to a mathematical treatment, such has been provided, along with analytical homework problems. For those topics that are best treated qualitatively, a tutorial level presentation has been provided along with review homework problems.

In general, I have concentrated on the theory (in the best sense of the word) behind these areas. I have also assumed little prior knowledge on the part of the reader concerning network planning. The wide coverage of topics means that the reader will see only the "tip of the iceberg." Hopefully, the references will guide the reader to more detailed sources on the various areas.

Thomas G. Robertazzi
Stony Brook, New York

ACKNOWLEDGMENTS

I would like to express my thanks to many people who helped me in the course of this five-year, part-time project. I am grateful to IEEE Press for publishing this work. The editorial assistance of Karen Hawkins and Linda Matarazzo is greatly appreciated. The production efforts of Savoula Amanatidis have been superb. The clarity of the text has benefited from the copyediting of Brenda Griffing. My understanding of this material has been improved through teaching the Stony Brook students in my graduate course on network management and planning. I would like to thank S. Charcranoon and Drs. S. Slotnick, R. Stadler, K. W. Tang, and A. Zhukovsky for reading and/or discussing specific portions of this manuscript. This book was written using computer facilities supervised by B. Carlson, K. Kantawala, A. Levochkin, and C. Pandya. G. Johnson, Stony Brook's engineering librarian, was very helpful in securing related material. My work at Stony Brook has benefited in general from the excellent secretarial skills of J. Eimer, C. Huggins, M. Krause, and D. Marge. Certain figures were drawn by H.-Y. Huang.

Last, but certainly not least, I'd like to thank my wife, Marsha, and my daughters, Rachel and Deanna, for their tremendous support in helping me to get where I am today in writing this book.

Thomas G. Robertazzi
Stony Brook, New York

CHAPTER 1

THE NETWORK PLANNING PROBLEM

1.1 INTRODUCTION

We are living in the midst of an industrial and social revolution involving the generation, processing, and transmission of information. The topic of this volume is a particular aspect of the technology of information transmission called "network planning."

What is network planning? Information today is transported over a variety of "networks"—collections of communication links and communication nodes. These range from analog and digital wired telephone networks to cellular and personal communication networks to satellite networks to, finally, data networks such as the Internet, ethernets, token rings, ATM local area networks, and metropolitan area networks. A particular organization may run a network comprising several of these technologies.

Any organization possessing and connected to such networks would like to be able to rely on orderly deployment, use, and upgrade of the networking equipment (both hardware and software). This is where the "planning" comes in. The role of network planning (and planners) is to provide an intelligent means for an organization to meet future network needs, using the existing networks as a starting point.

Naturally, as time proceeds, new developments occur. That is why planning is an ongoing process that continually makes use of the latest information to revise plans. This temporal property is a key aspect of the field of network planning—it involves the orderly evolution of networks *over time*.

Network planning is most highly developed in large telecommunication companies, which usually have distinct network planning departments for distinct systems. The large investments involved and today's increasingly competitive environment not only makes efficient planning a necessity; they serve to justify the use of sophisticated planning techniques. More recently, however, organiza-

1

tions with substantial network installations such as private companies, academic institutions, and government agencies, have found some sort of planning function to be necessary.

1.2   AN OPTIMIZATION PROBLEM

The interested reader may recognize what is being described as an optimization problem. Generically, one would like to minimize the cost of outlays for equipment and operations, maximize revenues if one is considering a profit-oriented organization, and do all this *over time* as technology, user requirements, and the economic environment change. Certainly this is a tall order.

In fact, it is impossible to pose the network planning problem, in all its baroque detail, as a single optimization problem for any reasonably sized organization.

Why? There are many reasons. One is the absence of a single optimization criterion. Certainly, cost, reliability, public image, capacity, and potential for growth are all important criteria, but how does one decide to balance them? Another reason is the size of the problem in large organizations, no matter whether it is measured in terms of the number of constraints, the number of people involved, and/or the number of organizational units involved. A third reason is the role of the unexpected: free market competition, new technological developments, and new economic developments.

Therefore what happens in reality is that the overall problem is divided into a (possibly large) number of smaller, more manageable subproblems. The partitioning of the subproblems may not be optimal, and only some of the subproblems may possess optimal solutions. Still, as any good scientist or engineer knows, only by abstracting out the essentials of a situation and posing the problem in a tractable manner can progress be made.

In fact the network planning problem is usually divided temporally as well into short-term, medium-term and long-term plans. These plans, naturally, should be consistent with one another. Moreover they must be periodically updated in what Gupta [Gupt] calls a "continual iterative process."

1.3   PLANNING FACTORS

The context in which a network planner operates is influenced by a number of factors. According to King and Premkumar [King], these include technology factors, business factors, organizational factors, and environmental factors. We discuss each set of factors in turn.

1.3.1 Technology Factors

A key property of technology is the periodic development of new technology that, eventually if not initially, can provide more capacity, increased reliability, and/or reduced cost. The network provider, whether a common carrier or a private organization, must then determine when and how to introduce this technology into an existing network.

Examples of the gradual displacement of older technologies include the use of digital technology in place of analog technology and the use of fiber optic links in place of electrical links. Some new technologies live alongside older technologies, as in the case of transoceanic satellite channels and undersea cables. Finally, when there is no existing infrastructure—as in large parts of today's world—there may be a technological solution such as wireless technology.

There are a plethora of issues that face the network planner as he or she considers a network upgrade. It is obviously necessary to ascertain the point at which the benefits of the upgrade outweigh the costs of deploying new technology. But there are other issues, such as the economic benefits of automation, the amount of importance attached by the network planner to integrating voice, video, and data over a single network, and the perceived desirability of open systems and standards [King].

1.3.2 Business Factors

Let's consider a common carrier service provider first. A key business factor is access to capital for building network infrastructure. A good example of this is the necessity to secure financing by LEO (low earth orbit) satellite providers such as Motorola's Iridium system. While such LEO systems have a certain technological glamour attached to them, the need to pay for infrastructure development applies equally to more mundane telecommunication systems.

There are a number of issues faced by a network planner at a common carrier. In many cases the cost of networks must be traded against their reliability. For instance, the widespread introduction of fiber optics has meant that a network may be able to function with fewer transcontinental links, thus saving on construction costs. However, the failure of one of the small number of remaining links can be catastrophic.

The introduction of new technology, once financed, can lower costs and increase reliability. However retiring old equipment may present a loss if an extended time in service had been contemplated.

Free market competition presents a wild card that can upset the best laid plans. The uncertainty in user demand for new services and the cost of deploying new services can present a situation that is akin to asking what came first—the chicken or the egg? This last problem gives weight to the argument for integrated network technologies such as ATM. A truly integrated network could support new services without the expense of dedicated specialized networks.

A common carrier network planner also faces issues of access to markets, spectrum allocation for wireless services, market penetration, and revenue maximization.

A user organization faces many of the same issues as a common carrier in terms of trading cost against reliability and the introduction of new technology. A user organization in today's business environment may have many common carriers and vendors to choose from for services and equipment. Determining the "best" offering can be a significant task in itself.

Finally, for user organizations that are profit driven, King and Premkumar [King] have stated that various strategies for gaining competitive advantage can be used in conjunction with properly planned telecommunications. These strategies include product differentiation, internal cost reduction, improved accessibility to markets, spawning new businesses, and changing competitive scope.

1.3.3 ORGANIZATIONAL FACTORS

A great deal has been written over the years on the best ways in which organizations should operate. This literature is not recapped here. However we make a few specific points for common carrier service providers.

Perhaps not in the short term, but in the long term, organizations must exploit new technologies to survive and prosper. Thus it is important to guard against the development of a "not invented here" culture in a common carrier organization. The skepticism with which satellite communication and packet switching were greeted with by some in the telecommunications industry is an example of this.

More recently, with the increased competition in the telecommunications industry, one may wonder whether telecommunications is driven by technology or cost. A technology-driven organization can be expected to allocate substantial resources for research and developments efforts. A cost-driven organization may view telecommunications as a commodity that must be brought to the public, organizationally, at the lowest possible cost. Naturally the telecommunications world is more complex than this. Successful research and development can reduce costs and, in the drive to reduce costs, it is sometimes necessary to use new technologies. Still, an organization's mind-set can influence the decisions that are made.

For a user organization [King], there are often noted similarities between information (computer) processing and telecommunications activities. This similarity has led to efforts at integration—both technologically and organizationally. In any case, the network planner at a user organization must be knowledgeable about a broad mix of technologies. Resources (i.e., the amount of money available for network operations and upgrades) form a key constraint in the duties of such a planner. Two important concerns of network planners in user organizations are the visibility of networking in the organization and the involvement of top managers in key decision making.

1.3.4 Environmental Factors

In this context, environmental factors are taken to be regulatory, economic and health questions. The environment a network planner at a common carrier service provider operates in is quite complex. The regulatory environment varies from country to country. The spectrum for wireless services in particular is heavily regulated. Attempts to "deregulate" telecommunications in recent years have led to a regulatory environment that is simpler, but more competitive. In the United States it is expected that telephone and cable companies will be competing for what was traditionally viewed as each other's core business.

The economic environment adds another layer of complexity to the network planner's role. The economic environment affects the resources available for network operation and expansion, the cost of money, and the money available for customers to spend on telecommunications. The economic environment varies from country to country, and its future evolution is hard to predict. Naturally the economic environment is linked to the competitive environment.

Finally, there can be public health concerns such as those recently raised concerning wireless technology and high frequency electromagnetic fields.

Many of these same issues affect—either directly or indirectly—user organizations, where the economic environment appears in the form of budgetary concerns. Transborder information flow regulations cause concern in multinational organizations. Standards are important for user organizations desiring multiple vendor sources for critical equipment.

1.4   TYPES OF PLANNING

There are a variety of classification approaches for different types of network planning. We partially follow the authors already cited [Gupt] [King] in listing a number of possibilities for common carrier service providers and user organizations.

1.4.1 By Level of Detail

- **Administrative planning** Goals, long-range plans, financial policy, regulatory efforts, forecasting, technology trends, strategies for competitive advantage
- **Fundamental technical planning** Plans for network, management, switching and routing, addressing, signaling, operations, provisioning, and maintenance
- **Engineering** Detailed and immediate plans

1.4.2 By Network Components (Telephone Nets)

- Local exchanges
- Toll exchanges
- Interexchange transmission
- Loop plant
- Signaling network
- Customer premises equipment

1.4.3 By Network Components (Computer Nets)

- Local area networks
- Packet networks
- PCs, workstations, minis, mainframes
- Routers, bridges, gateways
- Links (cables and wires, fiber optics, radio)

1.4.4 By Network Services (Telephone Nets)

- Plain old telephone service (POTS)
- Narrowband ISDN service
- Broadband ISDN service
- SMDS and/or frame relay service
- Packet service
- Video service
- Cellular telephone service

1.4.5 By Network Services (Computer Nets)

- E-mail
- Remote login
- File transfer
- Image transfer
- Voice connection(s)
- World Wide Web

1.4.6 By Time

- Long term (5–20 years)
- Medium term (2–5 years)
- Short term (1–2 years)

A temporal breakdown of ways to classify plans for a user organization appears in [King] [Clel].

1.5   NETWORK FEATURES

Typical communication networks have a number of characteristics that are independent of the actual implementation. These include [Gupta] the following.

1.5.1 Statistical Loads

Demand for telecommunication services is the result of the decisions of perhaps millions of individual users as to the time at which a service will be engaged, the type of service, and its duration. The time scales of these activities varies widely [Gupt], from milliseconds for e-mail to minutes for a voice call to weeks for a leased line. There are daily variations, weekly variations, and annual variations in traffic. Certain unpredictable events such as earthquakes or storms can also stimulate significant demand.

There is a large body of literature available on the characterization and prediction of this demand using statistical techniques. Among the most relevant techniques are stochastic processes and queueing theory [Gros] [Klei 75] [Robe 94]. Stochastic processes usually deal with the evaluation of series of events over time. Queueing theory, named for the British word for a waiting line, is the study of things waiting in a line. These things could be packets in a switch or calls in an exchange. Queuing theory is widely used to predict the performance of electronic systems.

1.5.2 Large Number of Subnetworks and Services

Most large "networks" are composed of a variety of subnetworks. This is particularly true of the Internet, which is an amalgamation of hundreds of smaller networks. But it is also true of large telephone networks. Here one has local networks, regional networks, long-distance transcontinental networks, and international networks. Some portions of these networks are segmented off into private corporate and government networks [Gupt].

Most existing large networks also offer a wide variety of services, some of which were outlined in Section 1.4. The large variety in subnetworks and services requires dedicated and specialized planning.

1.5.3 Growth

While POTS (plain old telephone service) is a relatively mature technology with low growth rates in developed countries, there is the potential for rapid growth in underdeveloped countries. Moreover, while a market may be static in overall growth, individual service providers and geographic areas may experience large growth rates in the process of attracting new customers and residents, respectively.

Certain parts of the telecommunications market have experienced rapid growth in recent years. This is particularly true of wireless technology such as cellular telephones. Wireless technology has experienced rapid growth in developed countries because of its convenience and in underdeveloped countries because of the need for minimal infrastructure. During the 1980s the number of local area networks deployed grew rapidly. It remains to be seen whether ATM technology will follow this pattern. Internet traffic has experienced phenomenal growth over the past decade with no end in sight.

Rapid growth in networks makes it necessary to install higher capacity systems and is an important motivation for proper network planning.

1.5.4 Hardware and Software Variety and Advance

A large network has a large variety of installed hardware and software with differing dates of manufacture. This is because of technological change (which makes new equipment cost effective) and the large investments involved (which lead to the use of equipment for extended periods of time). Naturally, this diversity of hardware and software, while economical, complicates network management and planning.

1.5.5 Technological Change

New technologies are being deployed into networks at an increasing rate. The coexistence of competing technologies is a reason for caution on the part of the network planner. On the up side, new technologies can promise reduced operating costs, enhanced capabilities, and increased reliability. However, it may not make sense to invest in a new technology unless one is sure that it has staying power in terms of market share, can be supplied by multiple vendors, and really does offer some or all of the advantages cited above.

1.5.6 Investing in the Future

Ideally telecommunications network infrastructure should last 8 to 20 years before requiring replacement. Thus the choice of hardware and software requires careful consideration on the part of the network planner.

1.5.7 Standards

In the network planning process the role of standards is important. Open standards allow a variety of vendors to bring compatible hardware and software to market at low prices. However, the timing of standards is important. Standardization at too early a point may suffer from incomplete research and development experience. Standardization at too late a point may be irrelevant if large number of users are locked into proprietary products and standards. A successful standard gives assurance to the network planner that hardware and software will be readily available on a long-term basis.

1.5.8 Economies of Scale

The phenomenon of economies of scale is responsible for the tendency of larger systems to be more cost effective than smaller systems. That is, while a larger system will cost more than a smaller system in total investment, the rate of increase actually decreases as systems become larger. Thus, on a per-user or per-circuit or per-packet basis, the larger system is actually cheaper.

Naturally, this doesn't imply that organizationally, larger is necessarily better. But in terms of specific telecommunications systems, such as transmission facilities or switches, savings due to economies of scale can be quite real.

1.5.9 Finite Resources

The resources available to a network planner for network modernization are usually limited and in fact can be significantly less than ideal. This serves to encourage a conservative approach to network planning. Awareness that resources are finite also provides a powerful incentive to try to install the right amount and type of equipment at the right time and place. Finally, the presence of such limitations necessitates the involvement of upper management in the planning process.

1.6   THE REST OF THIS BOOK

This book gives a tutorial look at the fundamentals of the theory of network planning. Chapter 2 discusses the use of mathematical programming for planning. Mathematical programming is widely used in large telecommunications companies to reduce costs. Chapter 3 examines a variety of other algorithms for network planning. Reliability theory is discussed in Chapter 4. This theory is used in a variety of applications including network planning. While general text-length treatments are available, key concepts are discussed in Chapter 4. Chapter 5 introduces key software technologies as well as a number of new optimization methodologies useful for network planning. Finally, Chapter 6 covers data analysis techniques that are network planning related.

1.7   PROBLEMS

1 What is network planning?

2 Why is network planning called a "continual iterative process"?

3 Why can network planning be viewed as an optimization problem?

4 Why is it impossible to pose the network planning problem for any reasonably sized organization as a single optimization problem?

5 Why is the network planning problem divided into smaller subproblems?

6 What criteria must an organization balance through network planning?

7 Give some examples of technologies that have displaced older technologies.

8 What are some of the issues that face a network planner who is considering a network upgrade?

9 Why is new technology periodically developed?

10 What are some of the business factors faced by a planner at a common carrier? At a user organization?

11 How does the uncertainty in demand for new user services affect network planning?

12 What are some strategies for gaining competitive advantage for organizations that are profit driven?

13 Why must organizations exploit new technologies to survive and prosper?

14 What is the difference between a telecommunications organization that is cost driven and one that is technology driven?

15 What are some important concerns of network planners in user organizations?

16 Describe the operating environment of a network planner in a common carrier. A user organization.

17 Differentiate types of planning by level of detail.

18 How are types of planning for network services and for network components related?

19 Why are network loads considered statistically? What are some of the relevant techniques used?

20 Why does a large network have a multiplicity of subnetworks and services?

21 Can there be organizational growth in a static market?

22 Why is there usually a large variety in the base of installed hardware and software?

23 How does technological change relate to network planning?

24 How long should telecommunications infrastructure be expected to last before it must be replaced?

25 What are some of the trade-offs between introducing technical standards too early and too late?

26 What is the concept of economies of scale?

27 What are some of the effects of the limitations on resources available to a network planner for network modernization?

2

MATHEMATICAL PROGRAMMING FOR PLANNING

2.1 OPTIMAL SOLUTIONS

Different problems in science, business, engineering and, of course, network planning, can be formulated mathematically. This mathematical formulation consists of two parts. The first part is the **objective function**. This is a mathematical function that tells us, for any proposed solution, what the "cost" of that solution would be. While in business applications cost is usually measured in terms of money, another quantity related to performance or reliability also might be used. The variables that describe the possible solution are called **decision variables**.

The second part of the mathematical formulation being discussed is a set of **constraints** expressed as a set of mathematical equations. A constraint puts one or more limitations on the range of acceptable solutions. For instance, some quantities may be only positive; some quantities may have lower or upper limits, however, or the weighted sums of a number of quantities may be constrained in these ways. Taken together, the set of all the constraints determines the legitimate or **feasible solution space**. This is the collection of all solutions that "makes sense" in terms of the limitations expressed by the constraints.

An objective function and associated set of constraints is usually called a **mathematical program**. The first question one may ask, once a mathematical program has been formulated for a given problem, is whether there is an optimal solution. An **optimal solution** is one that minimizes, or in some cases maximizes,[1] the objective function and also satisfies the set of constraints.

Some mathematical programming problems have a single, unique optimal solution. Yet it can happen that a mathematical program has a single, unique **globally optimal** solution but also has many **locally optimal** solutions. A globally optimal solution is optimal with respect to the entire feasible solution

---

[1] For instance, it is usually desirable to minimize cost but maximize reliability.

space. On the other hand, a locally optimal solution is optimal with respect to only a limited portion of the feasible solution space. The existence of locally optimal solutions usually makes the search for a global optimal solution difficult.

Mathematical programs can often be solved by **algorithms**. These are solution procedures that consist of a number of understandable steps and can be implemented on a computer. Some algorithms for some problems are exact and will be guaranteed to always produce a globally optimal solution. Sometimes an algorithm to produce a globally optimal solution is not known, or the available algorithm(s) are too slow to deal with a problem of a certain size. **Heuristic algorithms**, which use intuitive procedures to develop solutions that may be close to being optimal, may be useful in such cases.

This chapter examines a very useful type of mathematical programming called **linear programming**. It and its cousins have long been used in network planning to optimize telecommunications network design. The sections that follow first describe the basic (canonical) statement of the linear programming problem. Next comes a description of ways in which this statement can be tailored to particular families of problems. This is followed by a discussion of solution techniques for linear programming and some special cases, such as integer linear programming.

2.2 LINEAR PROGRAMMING

2.2.1 The Problem Statement

Linear programming refers to a particular type of linear problem statement and associated solution techniques. The problem formulation and the **simplex** solution technique, developed originally by George Dantzig in 1947, can be found in a more recent book by the same author [Dant]. The approach has become one of the most widely used procedures in computing. The standard (canonical) form of the problem to be described consists of two parts. One is the objective function to be minimized (or maximized):

$$\min Z = c_1 x_1 + c_2 x_2 + c_3 x_3 + \cdots + c_n x_n \tag{2.1}$$

Here the $x_i (i = 1, 2, \ldots, n)$ are the **decision variables** [Hill 90] whose optimal value has to be found. The $c_i (i = 1, 2, \ldots, n)$ are the proportional costs that multiply each of the x_i. Here Z is the total cost. The c_i and x_i are, in this statement, continuous real numbers, though as seen below, both are usually nonnegative. Naturally if we have an algorithm that minimizes Z, then a maximization problem can be handled by using $-Z$ as the objective function.

The second part of the canonical form is a set of constraints. There are m linear constraint equations for the decision variables:

$$a_{11}x_1 + a_{12}x_2 + \cdots + a_{1n}x_n = b_1 \tag{2.2}$$
$$a_{21}x_1 + a_{22}x_2 + \cdots + a_{2n}x_n = b_2 \tag{2.3}$$
$$a_{m1}x_1 + a_{m2}x_2 + \cdots + a_{mn}x_n = b_m \tag{2.4}$$

Here the a_{ij} and b_i are constants. There are also constraints, in the usual network planning problems, of the form:

$$x_1 \geq 0.0 \quad x_2 \geq 0.0 \quad x_n \geq 0.0 \tag{2.5}$$

Note that both the objective function and the constraint equations are **linear functions** of the $x_i (i = 1, 2, \ldots, n)$. This accounts for the "linear" in the phrase linear programming. The "programming" refers more to the planning aspects of the problem than to actual computer programs—though linear programs are almost always solved on a computer. As one might guess, there is a field called **nonlinear programming** dealing with the solution of nonlinear models.

The canonical form of linear programming just described can be written in vector notation [Luen 73]. If one defines

$$\mathbf{c} = \begin{bmatrix} c_1 \\ c_2 \\ \vdots \\ c_n \end{bmatrix} \quad \mathbf{x} = \begin{bmatrix} x_1 \\ x_2 \\ \vdots \\ x_n \end{bmatrix} \tag{2.6}$$

then the objective function, in matrix terminology, is

$$\min Z = \mathbf{c}^{\mathrm{T}}\mathbf{x} \tag{2.7}$$

The constraint equations can be written as

$$\mathbf{Ax} = \mathbf{b} \tag{2.8}$$

Here $x \geq 0$ on an element-by-element basis. The matrices $\mathbf{A}$ and $\mathbf{b}$ are

$$\mathbf{A} = \begin{bmatrix} a_{11} & a_{12} & \cdots & a_{1n} \\ a_{21} & a_{22} & \cdots & a_{2n} \\ \vdots & \vdots & \vdots & \vdots \\ a_{m1} & a_{m2} & \cdots & a_{mn} \end{bmatrix} \quad \mathbf{b} = \begin{bmatrix} b_1 \\ b_2 \\ \vdots \\ b_m \end{bmatrix} \tag{2.9}$$

2.2.2 SOME LINEAR PROGRAMMING EXTENSIONS

Two extensions that further the usefulness of the canonical linear programming formulation of the preceding section are described below. The first extension

[Luen 73] allows one to include linear constraint equations that are inequalities, rather than equalities. For instance, consider the following linear program,

$$\min Z = c_1 x_1 + c_2 x_2 + c_3 x_3 + \cdots + c_n x_n \tag{2.10}$$

subject to

$$a_{11} x_1 + a_{12} x_2 + \cdots + a_{1n} x_n \leq b_1 \tag{2.11}$$
$$a_{21} x_1 + a_{22} x_2 + \cdots + a_{2n} x_n \leq b_2 \tag{2.12}$$
$$a_{m1} x_1 + a_{m2} x_2 + \cdots + a_{mn} x_n \leq b_m \tag{2.13}$$

and also

$$x_1 \geq 0 \qquad x_2 \geq 0 \qquad x_n \geq 0 \tag{2.14}$$

Now introduce a number of nonnegative variables $y_i (i = 1, 2, \ldots, m)$. These y_i are called **slack variables**. The linear program now becomes

$$\min Z = c_1 x_1 + c_2 x_2 + c_3 x_3 + \cdots + c_n x_n \tag{2.15}$$

subject to

$$a_{11} x_1 + a_{12} x_2 + \cdots + a_{1n} x_n + y_1 = b_1 \tag{2.16}$$
$$a_{21} x_1 + a_{22} x_2 + \cdots + a_{2n} x_n + y_2 = b_2 \tag{2.17}$$
$$a_{m1} x_1 + a_{m2} x_2 + \cdots + a_{mn} x_n + y_m = b_m \tag{2.18}$$

and also

$$x_1 \geq 0 \qquad x_2 \geq 0 \qquad x_n \geq 0 \tag{2.19}$$

$$y_1 \geq 0 \qquad y_2 \geq 0 \qquad y_m \geq 0 \tag{2.20}$$

The y_i can be seen to be called slack variables because they make up the (slack) difference between the left side of the ith constraint and b_i. A linear program would solve for both the x_i and y_i but, naturally, only the solution of the x_i would be of major interest.

The linear program with slack variables can be put in the matrix equation form [Luen 73],

$$\min Z = \mathbf{c}^T \mathbf{x} \tag{2.21}$$

with

$$[\mathbf{A} \ \mathbf{I}] \begin{bmatrix} \mathbf{x} \\ \mathbf{y} \end{bmatrix} = \mathbf{b} \tag{2.22}$$

and also

$$\mathbf{x} \geq 0 \qquad \mathbf{y} \geq 0 \tag{2.23}$$

On the other hand, if one has an inequality of the form

$$a_{i1}x_1 + a_{i2}x_2 + \cdots + a_{in}x_n \geq b_i \qquad (2.24)$$

one can use a **surplus variable**

$$a_{i1}x_1 + a_{i2}x_2 + \cdots + a_{in}x_n - y_i = b_i \qquad (2.25)$$

where

$$y_i \geq 0 \qquad (2.26)$$

Here the term "surplus variable" accounts for the positive difference between the left side of the constraint equation and b_i. Naturally a linear program may make use of both slack and surplus variables along with some plain equality constraints.

The second extension [Luen 73] to the usefulness of linear programming that should be discussed involves **free variables**. These are x_i that have the range $-\infty < x_i < \infty$. Two methods for dealing with this situation will be given.

METHOD 1
Consider the example linear program

$$\min Z = 5x_1 + 2x_2 + 3x_3 + 1x_4 \qquad (2.27)$$

with

$$2x_1 + 3x_2 + 2x_3 + 5x_4 = 10 \qquad (2.28)$$
$$4x_1 + 1x_2 + 1x_3 + 7x_4 = 3 \qquad (2.29)$$
$$7x_1 + 2x_2 + 3x_3 + 1x_4 = 2 \qquad (2.30)$$

$$x_1 \geq 0 \qquad x_3 \geq 0 \qquad x_4 \geq 0 \qquad (2.31)$$

Here x_2 is the free variable. From the second constraint this can be expressed as

$$x_2 = 3 - 4x_1 - x_3 - 7x_4 \qquad (2.32)$$

This can be substituted into the objective function and also into the other two constraints to obtain a modified linear program:

$$\min Z = -3x_1 + x_3 - 13x_4 \qquad (2.33)$$

with

$$-10x_1 - 1x_3 - 16x_4 = 1 \qquad (2.34)$$
$$-1x_1 + 1x_3 - 13x_4 = -4 \qquad (2.35)$$

Note that the constant "6" was dropped in the objective function, since its absence will not affect the optimization.   ■

METHOD 2

In this method two new variables are introduced:

$$v \geq 0 \qquad w \geq 0 \tag{2.36}$$

Suppose that $-\infty < x_1 < \infty$. Then let

$$x_1 = v_1 - w_1 \tag{2.37}$$

In this way v_1 and w_1 will be positive and x_1 can be either positive or negative.

In the mathematical program one substitutes $v_1 - w_1$ for x_1. The program is now solved over the variables:

$$v_1, w_1, x_2, x_3, \ldots, x_n \tag{2.38}$$

∎

2.3   EXAMPLES OF LINEAR PROGRAMMING PROBLEMS

2.3.1 Introduction

This section describes a number of special cases of linear programming problems. It makes sense to identify these not only for the sake of obtaining insight into the types of applied problems to which linear programming can be applied but because a special mathematical structure is associated with each of these problems. Moreover special algorithms are sometimes available for the solution of these special cases. Finally, several of these problems can be transformed into one another. Specifically, it will be shown how each of the problems discussed can be converted into the **min cost flow problem**. An earlier listing of such transformations appears in [Hill 90]. While special-purpose algorithms exist for these problems, the transformations allow the use of general linear programming algorithms such as the simplex method or interior point algorithms.

2.3.2 The Min Cost Flow Problem

In the min cost flow problem [Luen 73] there are N nodes. A **node** is an entity, such as a computer, that is capable of processing data and can be connected by **arcs** (really links) to other computers so that data can be exchanged. In this problem a "product" of some sort (i.e., data) is assumed to "flow" between the nodes. Each node is either a source or a sink for this flow. The amount of flow originating at or terminating from a node is:

$$b_i > 0 \text{ source} \tag{2.39}$$

$$b_i < 0 \text{ sink} \tag{2.40}$$

Naturally, all of the product being generated should be consumed:

$$\sum_{i=1}^{N} b_i = 0 \qquad (2.41)$$

The cost of a unit of flow from node i to node j is, in the usual notation, c_{ij}. The amount of flow between node i and node j is x_{ij}. The mathematical program then becomes

$$\min Z = \sum_{i,j} c_{ij} x_{ij} \qquad (2.42)$$

with the constraints

$$\sum_{j=1}^{N} x_{ij} - \sum_{k=1}^{N} x_{ki} = b_i \qquad i = 1, 2, \ldots, N \qquad (2.43)$$

$$x_{ij} \geq 0 \qquad i, j = 1, 2, \ldots, N \qquad (2.44)$$

Note that the first set of constraint equations expresses the difference between the flow out of a node and the flow into the same node as b_i. Naturally equation (2.41) is not part of the mathematical program, but the b_i must be chosen to satisfy it. An inspection of the min cost flow problem will show that it can be solved by linear programming. This is in fact true of all the problems in this section.

Network Planning Applications.   We could use the min cost flow program to optimize the flow of data from a set of database machines (sources) to users (sinks). The cost may represent such factors as the distance between each database machine and each user, the cost of communications on particular links, and the cost of service from each database machine. Note that in this problem there are no constraints on the amount of flow from each source to sink.

2.3.3 The Transportation Problem

In this problem [Hill 90] [Luen 73] quantities of a single product, $a_1, a_2, \ldots, a_M$, are transported from M locations. The single product is received at N destinations in quantities $b_1, b_2, \ldots, b_N$. The cost of transportation for a unit of product is c_{ij}. The amount of flow between node i and node j is x_{ij}. Naturally, the objective is to minimize the total transportation cost:

$$\min Z = \sum_{i=1}^{M} \sum_{j=1}^{N} c_{ij} x_{ij} = \sum_{i,j} c_{ij} x_{ij} \qquad (2.45)$$

As one might expect, there are two types of constraint in the transportation problem. One holds that the sum of the product leaving the ith source is a_i. The second says that the sum of product received by the jth destination is b_j. Thus:

$$\sum_{j=1}^{N} x_{ij} = a_i \qquad \text{for } i = 1, 2, \ldots, M \tag{2.46}$$

$$\sum_{i=1}^{M} x_{ij} = b_j \qquad \text{for } j = 1, 2, \ldots, N \tag{2.47}$$

Naturally,

$$x_{ij} \geq 0 \text{ for } \begin{cases} i = 1, 2, \ldots, M \\ j = 1, 2, \ldots, N \end{cases} \tag{2.48}$$

Note that in this problem sources and sinks form two distinct subsets of nodes; that is, whether a node can supply product or receive it is predefined (unlike the rules governing the min cost flow problem).

There is also a natural conservation relation that holds that the sum of the product transported equals the sum of the product received:

$$\sum_{i=1}^{M} a_i = \sum_{j=1}^{N} b_j \tag{2.49}$$

This is not a constraint on the x_{ij} but must hold for all the a_i and b_i selected.

Network Planning Applications.   One could consider a telecommunications example in which circuits have to be connected from M switching centers, each of which has capacity $a_1, a_2, \ldots, a_M$, to N local sites in quantities $b_1, b_2, \ldots, b_N$. Again, the cost of connecting one circuit from switching center i to local site j is c_{ij}.

In this formulation we would be using the continuous x_{ij} to represent a quantity that should really only take on integer values (integer numbers of circuits). Using this approach to provide approximate answers can be problematic, as discussed below in Section 2.5 on integer programming. Of course, this is a basic model. More involved models may include such features as fixed charges per circuit, redundant paths, or a hierarchy of switching centers and circuit distribution.

Conversion to Min Cost Flow Problem.   An examination of the transportation problem and the min cost flow problem will show that they are very similar. A major difference in the formulations of this book is an artificial difference: the min cost problem uses signed b_i and the transportation problem uses positive a_i and b_i for supply and demand, respectively. It should be apparent that both cases result in the same numbers of equations, and one can convert easily from one form to another. A more germane difference is that a graph is usually associated with the transportation problem. In the graph arcs proceed from source nodes to sink nodes only. Each arc has no limit on its carrying capacity. The only other differences are ones of nomenclature. That is, in the min cost flow problem c_{ij} and x_{ij} are associated with the cost and the amount of flow. In the transportation problem c_{ij} and x_{ij} are associated with the cost and amount of transportation. In fact, of all the conversions between problems in this chapter, the one between the transportation problem and the min cost flow problem is the most straightforward.

2.3.4 The Transshipment Problem

The transshipment problem is similar to the transportation problem except that there are now a number of transshipment nodes in addition to the source and destination nodes. The transshipment nodes can be used to relay shipments of the "product" from sources to destinations. Let's break the set of nodes $[N]$ into three subsets, $[S, T, D]$. Here S represents the set of source nodes, T represents the set of transshipment nodes, and D represents the set of destination nodes. Again, let x_i be the amount of product shipped from node i to node j. In the context of network planning "product" can be data or circuits. Let c_{ij} be the cost of shipping a unit of product from node i to node j. Two forms of the transshipment problem are outlined. The first is

$$\min Z = \sum_{i,j} c_{ij} x_{ij} \qquad (2.50)$$

$$\sum_{j=1}^{N} x_{ij} - \sum_{k=1}^{N} x_{ki} = a_i \quad i \in S \qquad (2.51)$$

$$\sum_{j=1}^{N} x_{ij} - \sum_{k=1}^{N} x_{ki} = 0 \quad i \in T \qquad (2.52)$$

$$\sum_{j=1}^{N} x_{ij} - \sum_{k=1}^{N} x_{ki} = -b_i \quad i \in D \qquad (2.53)$$

Where $i \in S$ indicates that i is a member of set S. In this program there is a net outflow from each source node, a net inflow at each destination node, and a conservation of flow at each transshipment node. Also, a_i is the positive supply at the ith source node and b_i is the positive demand at the ith destination node. Naturally:

$$\sum_{i \in S} a_i = \sum_{i \in D} b_i \qquad (2.54)$$

In a **capacitated transshipment problem** there is also a set of constraints on the flow on each link:

$$0 \leq x_{ij} \leq \text{Cap}_{ij} \ \forall \ i, j \qquad (2.55)$$

Here Cap_{ij} is the maximum capacity on link ij. The notation $\forall\, ij$ means that all pairs of i and j are considered.

The second version of the transshipment problem is as follows [Ford]:

$$\min Z = \sum_{i,j} c_{ij} x_{ij} \tag{2.56}$$

$$\sum_{j=1}^{N} x_{ij} - \sum_{k=1}^{N} x_{ki} \le a_i \quad i \in S \tag{2.57}$$

$$\sum_{j=1}^{N} x_{ij} - \sum_{k=1}^{N} x_{ki} = 0 \quad i \in T \tag{2.58}$$

$$\sum_{j=1}^{N} x_{ij} - \sum_{k=1}^{N} x_{ki} \le -b_i \quad i \in D \tag{2.59}$$

In this case the a_i, b_i, are limits on the amount of product to be sourced and sinked, respectively. Now, rather than equation (2.54), we have the following constraint on the a_i and b_i:

$$\sum_{i \in S} a_i \ge \sum_{i \in D} b_i \tag{2.60}$$

Naturally, slack variables would be needed to put this linear program into the basic form. It is also possible to have a capacitated version of this program using the constraints in equation (2.55). One can see in this second problem that there is more flexibility in deciding which sources provide product and which destinations accept product. The appropriateness of either case depends on the application.

Network Planning Applications.   The transshipment problem is well suited to solving problems involving the shipment of data or voice circuits through intermediate (transshipment) nodes. A typical transshipment node that neither generates nor receives data/circuits might be a communications satellite, undersea cable, or terrestrial route (microwave, coaxial, or optical). Note that compared to the transportation problem, the transshipment problem includes the element of **routing** decisions.

Conversion to Min Cost Flow Problem.   It has been mentioned that the transshipment problem is a generalization of the transportation problem. From inspection of the min cost flow problem, it can be seen that the only significant difference with respect to the transshipment nodes is now the presence of constraint equations for the transshipment nodes where $b_i = 0$. Otherwise the conversion is straightforward.

2.3.5   The Assignment Problem

The canonical use of the assignment problem [Hill 90] [Luen 73] is to assign N workers to N jobs. Let the benefit (opposite of cost) of assigning worker i to job j be c_{ij}. The x_{ij} are

$$x_{ij} = \begin{cases} 1 & \text{if worker } i \text{ assigned to job } j \\ 0 & \text{otherwise} \end{cases} \tag{2.61}$$

Then the objective function, summing over all possible job assignments, is

$$\max Z = \sum_{i=1}^{N} \sum_{j=1}^{N} c_{ij} x_{ij} \tag{2.62}$$

The necessary constraints must express three facts about the problem.

- Each worker can be assigned to at most one job.
- Each job can be assigned only one worker.
- The x_{ij} are greater than or equal to zero.

This can be done as follows:

$$\sum_{j=1}^{N} x_{ij} = 1 \text{ for } i = 1, 2, \dots, N \tag{2.63}$$

$$\sum_{i=1}^{N} x_{ij} = 1 \text{ for } j = 1, 2, \dots, N \tag{2.64}$$

$$x_{ij} \geq 0 \text{ for } \begin{cases} i = 1, 2, \dots, N \\ j = 1, 2, \dots, N \end{cases} \tag{2.65}$$

As one might guess, this linear program only makes sense if $x_{ij} = 0$ or 1. That is, one can assign a particular worker to a particular job or not assign that worker to a particular job. Other choices of x_{ij} make no sense in this context. In fact, the use of a standard linear programming algorithm on this problem results in binary solutions. Effective algorithms to solve the assignment problem, other than linear programming, are available.

Network Planning Applications.   One can envision data and communications telecommunications applications such as assigning different high speed ports to different users or assigning long distance facilities to different users. The benefit, c_{ij}, in the case of a toll network would most likely be the revenue generated by the assignment. In the case of a "free" network, the network provider may wish to minimize the network's cost of providing free connections.

Conversion to Min Cost Flow Problems.   A two-step process can be used. One can show that the assignment problem is equivalent to the transportation problem, which was shown to be equivalent to the min cost flow problem. To show this first equivalence, one can let the number of source nodes in the transportation problem equal the number of sink nodes. Then $a_i = 1$ (in the transportation problem) if the ith node is a source node and $b_j = 1$ if the jth node is a sink node. This completes the conversion between the transportation problem and the assignment problem.

2.3.6 The Maximal Flow Problem

The maximal flow problem is a problem [Luen 73] with N nodes. One node is a source and one node is a sink. Label a source node "s" and a sink node "d" (for destination). There are also constraints on the capacity of each link:

$$0 \le x_{ij} \le Cap_{ij} \ \forall \ i, j \tag{2.66}$$

here Cap_{ij} is the capacity of the ijth link. The amount of flow between node i and node j is x_{ij}. The objective in this problem is to maximize the flow through the network (the usual c_{ij} are set to 0). So letting the flow from s to d be f, we have

$$\max f \tag{2.67}$$

$$\sum_{i=1}^{N} x_{si} - \sum_{i=1}^{N} x_{is} = f \quad \text{source} \tag{2.68}$$

$$\sum_{j=1}^{N} x_{ij} - \sum_{j=1}^{N} x_{ji} = 0 \quad i \ne \text{source, sink} \tag{2.69}$$

$$\sum_{i=1}^{N} x_{di} - \sum_{i=1}^{N} x_{id} = -f \quad \text{sink} \tag{2.70}$$

where f corresponds to flow arcs either leaving the source or entering the sink. Although the maximal flow problem can be solved by linear programming, special-purpose algorithms are available [Hill 90][Luen 73].

Network Planning Applications.   Direct applications include situations in which the number of circuits routed from A to B (source to sink) must be maximized while not exceeding the "free" circuit capacity of various links. A similar situation may arise in data transport.

Conversion to Min Cost Flow Problem.   It has been mentioned that the c_{ij} should be set to zero. As discussed in [Hill 90], one must also choose a safe upper bound on the maximal flow and set the potential supply at the source and the potential demand at the sink equal to this. Finally, to force the minimal cost flow algorithm to route a maximal flow through the original network, one adds a link of very high capacity and very high cost from source to sink.

2.4 SOLVING LINEAR PROGRAMS

There are two general approaches for solving linear programs. The first is the **simplex method** of George Dantzig [Dant]. The second is an **interior point algorithm** developed by Narendra Karmarkar of AT&T in 1984. To understand both approaches, consider the following linear program:

$$\max Z = x_1 + 2x_2 \tag{2.71}$$

$$\frac{4}{3}x_1 + x_2 \leq 4 \qquad \text{at } a \tag{2.72}$$

$$x_2 \leq 3 \qquad \text{at } b \tag{2.73}$$

$$4x_1 + x_2 \leq 8 \qquad \text{at } c \tag{2.74}$$

where a, b, and c are as shown in Fig. 2.1. If one plots the constraints, as in Figure 2.1, it is apparent that the feasible solution space is the convex region indicated. In fact the feasible solution space for a linear program is always a convex region with a dimensionality equal to the number of decision variables. If one thinks of this convex polyhedron as a diamond with a faceted surface, it turns out that the location of the optimal solution (if it is unique) is always exactly on a surface corner (as I, II, III, IV in Fig. 2.1). This is proven in any text on linear programming [Hill 90] [Leun 73]. These corner points are known as **extreme**

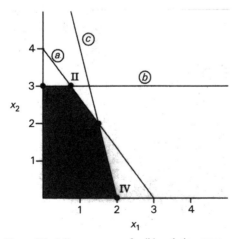

Figure 2.1 A linear program feasible solution space.

points. For the preceding linear program, the value of the objective function at the extreme points is

$$(0) + 2(3) = 6.00 \text{ at I} \tag{2.75}$$

$$\tfrac{3}{4} + 2(3) = 6.75 \text{ at II} \tag{2.76}$$

$$\tfrac{3}{2} + 2(2) = 5.50 \text{ at III} \tag{2.77}$$

$$2 + 2(0) = 2.00 \text{ at IV} \tag{2.78}$$

It can be seen that the optimal solution occurs at extreme point II. Dantzig's simplex method is a procedure in which the simplex algorithm moves from a consideration of extreme point to extreme point—in a directed, intelligent fashion—until the optimal extreme point is reached. Karmarkar's interior point algorithm starts from a position within the simplex and moves, in a step-by-step descent fashion, to the optimal extreme point.

2.5   INTEGER LINEAR PROGRAMMING

2.5.1 Introduction

Many a practical network planning problem, when formulated as a mathematical program, has a solution where some or all decision variables must be integers rather than continuous variables. If all the decision variables have integer solutions, such a program is known as an **integer linear program**. If only some of the decision variables have integer solutions, the term **mixed integer linear program** is applied.

To provide approximate answers to such problems, one might be tempted to use standard linear programming and then round the continuous answers to integer values. However there are problems with this approach. The resulting answers may not be optimal, or worse, even feasible. Let us consider two examples of integer linear programming problems, modified appropriately, from [Coop 74].

2.5.2 Example 1: Capital Budgeting

Suppose a telecommunications company is to build a series of N switching centers to provide switched service to a customer base. Let c_i be the cost of building a switching center at site i. Let p_i be the profit to be generated by having an operational switching center at site i. There is a constraint on the total amount of funds available to build switching centers,

$$\sum_{i=1}^{N} c_i x_i \leq C \tag{2.79}$$

where the decision variables, the x_i are defined as

$$x_i = \begin{cases} 1 & \text{if site } i \text{ is selected as a center} \\ 0 & \text{if site } i \text{ is not selected as a center} \end{cases} \tag{2.80}$$

Then

$$\text{total cost} = \sum_{i=1}^{N} c_i x_i \tag{2.81}$$

$$\text{total profit} = \sum_{i=1}^{N} p_i x_i \tag{2.82}$$

The mathematical program then becomes

$$\max Z = \sum_{i=1}^{N} p_i x_i \tag{2.83}$$

with the constraints

$$\sum_{i=1}^{N} c_i x_i \leq C \tag{2.84}$$

$$x_i \geq 0 \quad i = 1, 2, \ldots, N \tag{2.85}$$

$$x_i \text{ is a binary integer} \quad i = 1, 2, \ldots, N \tag{2.86}$$

2.5.3 Example 2: Fixed Charge Problem

Consider now a telecommunications services or computer services company that wants to offer a subset of a number of potential new services to its customers. Let x_i be the amount of service i offered and let c_i be the proportional cost of offering the service. Let p_i the proportional profit in offering an amount x_i of service i. Then:

$$\text{total profit} = \sum_{i=1}^{N} p_i x_i \tag{2.87}$$

The new wrinkle here is to define a **fixed charge**, k_i, that is a cost incurred by the company if service i is offered, independent of how much demand there is for the service. This might represent such fixed costs as establishing and maintaining the infrastructure for the service and initial advertising outlays. It is natural then to define the decision variable y_i,

$$y_i = \begin{cases} 1 & \text{if } x_i > 0 \\ 0 & \text{if } x_i = 0 \end{cases} \tag{2.88}$$

so that

$$\text{total cost} = \sum_{i=1}^{N} (c_i x_i + k_i y_i) \tag{2.89}$$

where x_i is a continuous variables and the y_i are integer variables. Assuming that there is some maximum amount, C, that may be invested in offering new services, one has as a mathematical program:

$$\max Z = \sum_{i=1}^{N} p_i x_i \tag{2.90}$$

The constraints are

$$\sum_{i=1}^{N} (c_i x_i + k_i y_i) \leq C \tag{2.91}$$

$$x_i > 0 \quad i = 1, 2, \ldots, N \tag{2.92}$$

$$x_i \text{ is a continuous variable} \quad i = 1, 2, \ldots, N \tag{2.93}$$

$$y_i \text{ is a binary integer variable} \quad i = 1, 2, \ldots, N \tag{2.94}$$

Since some of the decision variables are continuous and some are integers, this program is referred to as a mixed integer linear program. The problem is in fact difficult to solve because of the nonlinearity introduced by the presence of the y_i.

2.5.4 Solution Techniques

With N binary (0 or 1) integer variables, the number of potential solutions is simply 2^N. For example, with $N = 100$ there are 1.26×10^{30} solutions. Even a computer that could evaluate a billion solutions each second would take over 40,000 billion years to exhaustively evaluate all possible solutions! This is a basic problem in algorithm design. There is a large family of related and fundamental computational problems for which the computation beyond a certain problem size becomes intractable. These problems are said to be **NP complete**.

But what of integer programming? The inclusion of integer variables means that the optimal solution can no longer be guaranteed to be on an extreme point of the simplex. Thus the simplex algorithm cannot be used directly. In linear programming relaxation algorithms though, standard linear programming is used in parts of the algorithm. Heuristic algorithms—that is, ones based on some intuition regarding the problem—often produce solutions that are close to optimal. In fact, the min cost flow problem and its variations always have integer solutions if the b_i and arc capacities are integers. They can be solved by special versions of the simplex algorithm.

A popular approach to solving integer linear programs is the family of **branch and bound** algorithms. To understand how they work, one must understand the concept of a **decision tree**. A decision tree is a representation of all possible decisions to be made in producing all the possible solutions to a problem. Each potential solution can be thought of as occupying a node on the tree, and related solutions occupy subtrees within the tree. A branch and bound algorithm will, through some insight to a problem, develop bounds on the solution. These usually can be used to eliminate particular subtrees from consideration in the search for the optimal solution. Thus, if large portions of the decision tree are continually removed from consideration, one need not evaluate every possible solution but only a limited subset of solutions. This can drastically decrease the solution time.

2.6 DYNAMIC PROGRAMMING

The powerful approach to solving certain optimization problems known as **dynamic programming** was developed by R. Bellman [Bell 57] [Bell 62]. The basic idea is that a large family of optimization problems can be transformed into the problem of finding the shortest path through an appropriate graph. Shortest path algorithms are quite developed and thus the overall procedure is usually quite computationally efficient. As an example [Zade], consider a link in a telephone system for which the predicted demand for circuits is illustrated in Figure 2.2. Furthermore, assume that the systems in Table 2.1, with the following capacities and total costs, can be purchased.

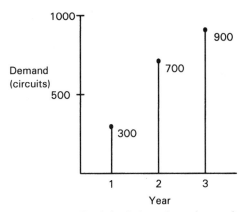

Figure 2.2 Predicted circuit demand over the next three years.

Table 2.1 Capacity and Cost

System	Capacity (circuits)	Cost
A	100	$1000
B	250	$1750
C	500	$3000

Note that on a per-circuit basis, the larger a system's capacity the lower the cost. This important relation can result in **economies of scale**. Now the problem is to decide how much capacity to install in each year to meet the predicted demand and minimize the total investment over three years.

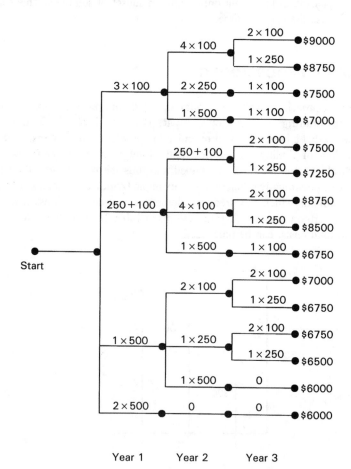

Year 1   Year 2   Year 3

Figure 2.3 Installation sequence decision tree.

The problem can be solved in the following manner. We start by creating a graph, or decision tree, illustrating each possible installation sequence over the three years (see Fig. 2.3). Moving from left to right, each branch represents a different decision that can be made in a particular year. Associated with each branch is a cost. A path from source node "0" to a node in year 3 represents a particular three-year installation sequence. The total cost of each sequence in Figure 2.3 appears at the right of the figure.

It can be seen that the worst strategy (costing as much as $9000) is to install in each year just enough capacity to meet demand. The best strategy is to install two 500-ciruit systems, though this can be done in two ways (2 in year 1 or 1 in year 1 and 1 in year 2). Naturally, running a shortest path algorithm through the graph would expose these optimal solutions.

One factor that could also be included in this example is the **cost of money**. That is, the cost of securing equipment may vary from year to year because of changes in the rate of inflation, return on investment and other factors. Thus, for example, the purchase of a 500-circuit system may be cheaper in some years than others. To include the cost of money in Figure 2.3, one simply makes the appropriate adjustment in the cost associated with each branch.

2.7   CASE STUDY I: A CELL ASSIGNMENT PROBLEM

One of the bright spots in electrical technology over the past several years has been the rapid growth of wireless products and services. In particular, mobile communications has evolved into an increasingly widely deployed technology.

The basic system architecture of mobile communication systems is to divide the geographic area to be covered into a collection of smaller areas or "cells." A radio base station at the center of each cell maintains radio contact with mobiles in the cell. The base station is also connected to a switching computer or "switch," which allows it to connect mobile calls to and from the public switched network. General treatments of mobile communication systems appear in [Mehr] [Rapp] [Redl].

An important concept in cellular communication systems is that of the "handoff." A handoff occurs when a moving mobile leaves one cell and enters an adjacent one. Then the mobile switches radio communications with the base station of the original cell to radio communications with the base station of the new cell. Thus we have a "handoff" of the mobile. This handoff is accomplished in a manner that is transparent to the mobile user. Essentially, radio signal strength is measured so that it is clear when a mobile is moving away from the coverage area of one base station and into the coverage area of another base station.

Future personal communication services (PCS) will make possible universal handheld communication devices. While the cells in such PCS systems will be smaller than the cells in current mobile communication systems, many of the same concepts apply to systems of both types.

Generally, a single switch may be connected to multiple cells. Thus there are two types of handoff. In one, a mobile moves from one cell to an adjacent cell, both of which are under the control of the same switch. This is a relatively simple handoff inasmuch as it involves only the one switch and no location updates. In the second type of handoff, a mobile moves from one cell to an adjacent one, each of which is connected to a different switch. This leads to complications in terms of the need to update location databases, a complex protocol between the two switches and the continued routing of the call through the first switch for billing [Merc].

The optimization problem to be solved in this case study was formulated and solved by Merchant and Sengupta [Merc]. Their approach is followed closely in the example below.

In this optimization problem there are N cells and M switches. The basic problem is to assign cells to switches so as to minimize a cost function.

The following variables will be defined:

h_{ij} Cost per unit time of handoffs from cell i to j. This is proportional to the volume of handoffs between cell i and cell j.

c_{ik} The cost per time of cabling between cell i and switch k. This cost could be adjusted (for amortization, the cost of money, etc.).

λ_i Number of calls processed by cell i in unit time.

M_k Call processing capacity of switch k.

Next a mathematical program is set up in the form of an integer linear programming problem. A variable x_{ik} is defined such that:

$$x_{ik} = \begin{cases} 1 & \text{if cell } i \text{ assigned to switch } k \\ 0 & \text{otherwise} \end{cases} \tag{2.95}$$

There is a constraint on the call processing ability of each switch:

$$\sum_{i=1}^{N} \lambda_i x_{ik} \leq M_k \quad k = 1, 2, \dots, M \tag{2.96}$$

A second constraint forbids each cell to be assigned to more than one switch:

$$\sum_{k=1}^{M} x_{ik} = 1 \quad i = 1, 2, \dots, N \tag{2.97}$$

Now the objective function has two additive components. One is the cost of cabling switches to cells:

$$Z_{\text{cabling}} = \sum_{i=1}^{N} \sum_{k=1}^{M} c_{ik} x_{ik} \tag{2.98}$$

The individual c_{ik} may be proportional to distance and/or take into account local cabling costs.

The second cost component is the cost of handoffs occurring between two distinct switches. To account for this, first define z_{ijk}:

$$z_{ijk} = x_{ik} x_{jk} \qquad i, j = 1, 2, \ldots, N \qquad k = 1, 2, \ldots, M \tag{2.99}$$

One can see that z_{ijk} is unity if cell i and cell j are assigned to the same switch, k, and is zero otherwise. Next, define y_{ij}:

$$y_{ij} = \sum_{k=1}^{M} z_{ijk} \qquad i, j = 1, 2, \ldots, N \tag{2.100}$$

where y_{ij} is unity if cells i and j are assigned to the same switch. Now one can see that the cost of handoffs (again, per unit time) is:

$$Z_{\text{handoff}} = \sum_{i=1}^{N} \sum_{j=1}^{N} h_{ij}(1 - y_{ij}) \tag{2.101}$$

It can be seen that h_{ij}, the cost per unit time of handoffs between cell i and cell j, makes a contribution toward Z_{handoff} only if cell i and cell j are connected to different switches. The overall cost function is

$$Z = Z_{\text{cabling}} + Z_{\text{handoff}} \tag{2.102}$$

$$Z = \sum_{i=1}^{N} \sum_{k=1}^{M} c_{ik} x_{ik} + \sum_{i=1}^{N} \sum_{j=1}^{N} h_{ij}(1 - y_{ij}) \tag{2.103}$$

A problem with the formulation above, pointed out by Merchant and Sengupta, is that the constraint of equation (2.99) is not linear. The program can be converted into an integer linear program by using the following constraints in place of equation (2.99):

$$z_{ijk} \le x_{ik} \tag{2.104}$$

$$z_{ijk} \le x_{jk} \tag{2.105}$$

$$z_{ijk} \ge x_{ik} + x_{jk} - 1 \tag{2.106}$$

$$z_{ijk} \ge 0 \tag{2.107}$$

Here $i, j = 1, 2, \ldots, N$ and $k = 1, 2, \ldots, M$. It can be shown that the constraint of equation (2.99) implies the set of constraints above, which in turn implies equation (2.99). Thus we now have a standard integer linear program that can be solved by standard techniques [Hill 90].

This case study is a good example of applying mathematical programming techniques to a practical and current telecommunications problem. Merchant and Sengupta [Merc] go on to consider a dual homing version of the problem in which the amount of calls and the pattern of handoffs are different during two times of day. They also propose an efficient heuristic algorithm for this problem.

2.8 CASE STUDY II: CELLULAR ECONOMIC MODEL

2.8.1 Introduction

The case study of Section 2.7 addressed a specific optimization problem involving the optimal assignment of cells to switches to minimize cost. This assumes that all cell locations are known. A larger and more complex problem is the problem of the overall efficient design of a complete cellular system. The problem is complex because of the many factors that must be considered. These include system performance specifications, topography, radio propagation, traffic patterns, and the cost of system components. The last item includes the cost of base station equipment.

These factors are interrelated. With our current understanding of optimization theory and with current computer technology, it makes sense to break the problem into a series of subproblems arranged in a hierarchical fashion. In this case study the middle of three subproblems of an overall hierarchical optimization planning (HOP) approach is described. This particular formulation is due to Hao, Soong, Gunawan, Ong, Soh, and Li [Hao].

The first subproblem addressed by the HOP approach is to make an initial determination of an upper bound on the number of cells needed and on cell size. This is done using a well-known radio propagation model developed by Hata [Rapp] and some simple sizing rules. In the second subproblem, a mathematical program called the economic optimization model (EOM), described below, is used to make an optimal determination of the number of cells, cell size, and cell allocation. This formulation can be solved by means of the technique of simulated annealing. The third subproblem entails detailed planning and accurate cost estimation involving base station location and provisioning as well as channel assignment. At this point the cost can be accurately calculated.

This three-subproblem process can be repeated (iterated) to meet desired quality of service objectives.

2.8.2 Economic Optimization Model (EOM)

In the EOM the geographic area to be considered is divided into "grids" so small that many grids will constitute a cell. The main decision variable is

$$x_{ik} = \begin{cases} 1 & \text{if grid } i \text{ belongs to cell } k \\ 0 & \text{otherwise} \end{cases} \tag{2.108}$$

This is augmented with

$$Y_k = \begin{cases} 1 & \text{if there are grids in cell } k \\ 0 & \text{if cell } k \text{ is empty} \end{cases} \tag{2.109}$$

$$G_{i1} = \begin{cases} 1 & \text{urban structure in grid } i \\ 2 & \text{suburban structure in grid } i \\ 3 & \text{rural structure in grid } i \end{cases} \tag{2.110}$$

The last distinction is made because of the difference in radio propagation characteristics of urban, suburban, and rural areas. Note that if $Y_k = 0$, it is not necessary to place a base station in the kth cell.

The objective function to be minimized in the EOM is

$$Z = C_{so} + \sum_k Y_k (C_{cell} + C_a g_{bk} + C_t P_{tk}) \tag{2.111}$$

where C_{so} is the fixed cost for the switching office, which provides network connectivity for the base stations, including hardware and installation. If cell k consists of a nonzero number of grids, then the base station cost is included. The terms within the parentheses are C_{cell}, the cost of hardware and installation for the base station; $C_a g_{bk}$, the product of the antenna cost efficient and the base station antenna gain[2]; and $C_t P_{tk}$, the product of the transmitting power cost coefficient and the base station transmission power. Note that there are technological lower and upper bounds (LB and UB, respectively) $g_{LB} \leq g_{bk} \leq g_{UB}$ and $P_{LB} \leq P_{tk} \leq P_{UB}$. One wishes to minimize the sum of the three quantities in the parentheses.

As with any mathematical program there are a number of constraints in EOM. In the following n is the total number of cells and m is the total number of grids. One first has

$$P_{tk} + g_{bk} + g_{mk} - L_k(d_k) \geq P_{cell} \qquad k = 1, 2, \ldots, n \tag{2.112}$$

This constraint holds that the sum of base station power, base station gain, and mobile subscriber gain, less the propagation loss predicted by the Hata model (all

---

[2] "Gain" is a measure of the efficiency of an antenna. Here higher gain antennas cost more.

in decibels), meets a minimum system radio power level at the coverage boundary. The second constraint is

$$\sum_i x_{ik} G_{i2} \le C_{pt} \qquad k = 1, 2, \ldots, n \qquad (2.113)$$

where G_{i2} is the traffic density (number of subscribers per hour) in grid i. Thus this constraint ensures that the total amount of traffic in a cell is at or below the amount of traffic that the cell can handle, C_{pt}. Note that this depends on the number of available channels and the desired blocking probability, P_{block}.

The next three sets of constraints ensure that a number of basic relationships concerning grids are met.

The first of the three sets of constraints is

$$x_{ik} x_{i'k} (G_{i1} - G_{i'1}) = 0 \qquad (2.114)$$

$$k = 1, 2, \ldots, n \qquad i, i' = 1, 2, \ldots, m \qquad i \ne i'$$

This constraint is zero when two grids belonging to the same cell have the same structure, or it is zero if two grids belong to two different cells. That is, it ensures that all the grids belonging to the same cell are of the same type.

The second of this group of three sets of constraints is

$$\sum_k x_{ik} = 1 \qquad i = 1, 2, \ldots, m \qquad (2.115)$$

It can be seen that it holds that each grid belongs to exactly one cell.

The final of this group of three sets of constraints is

$$\pi d_k^2 = \sum_i (x_{ik} S_g) \qquad k = 1, 2, \ldots, n \qquad (2.116)$$

here d_k is the radius of cell k so that πd_k^2 is the geographic area of cell k. Also, S_g is the area of a single grid. This set of constraints simply holds that only enough grids to fit in the coverage area of a given cell are assigned to that cell.

Finally, one has two additional sets of constraints:

$$d_k (Y_k - 1) \ge 0 \qquad k = 1, 2, \ldots, n \qquad (2.117)$$

$$d_k > Y_k - 1 \qquad k = 1, 2, \ldots, n \qquad (2.118)$$

These maintain the relationship that cell radius, d_k, is positive when a cell has at least one grid ($Y_k = 1$) and is zero when a cell has no grids ($Y_k = 0$).

Like the case study of Section 2.7, the EOM model is a good example of a practical engineering problem cast as a mathematical program. However, it is a challenging problem to solve because of the large number of variables and constraints and their complex relationships. Hao et al. [Hao] use the optimization approach of simulated annealing for solving this hard combinatorial problem. Simulated annealing is discussed in detail in Chapter 5.

2.9 PROBLEMS

1 **a.** Draw the feasible solution space of the following linear program and find the optimal solution. Show all work:

$$\max Z = 5x_1 - 5x_2 \tag{2.119}$$

Subject to

$$x_1 \le 10 \tag{2.120}$$

$$2x_1 + 10x_2 \le 100 \tag{2.121}$$

$$4x_1 + 5x_2 \le 65 \tag{2.122}$$

b. Put this linear program into canonical form.

2 In the following x_2 is a free variable. Transform this program into standard form:

$$\min Z = 2x_1 + 3x_2 + 7x_3 . \tag{2.123}$$

Subject to

$$4x_1 + x_2 + 5x_3 = 4 \tag{2.124}$$

$$2x_1 + 10x_2 + 3x_3 = 14 \tag{2.125}$$

$$x_1 \ge 0 \tag{2.126}$$

$$x_3 \ge 0 \tag{2.127}$$

3 Write a mathematical program for a modified assignment problem where each worker can be assigned to q jobs (simultaneously).

4 Write an equation(s) to implement the situation described in the following statements. Define all variables used.

   a. The sum of all "products" (i.e., data, circuits, etc.) generated is greater than or equal to the sum of the product consumed over an entire network.

   b. Minimize the cost of transporting product between all node pairs i and j in a network.

   c. A decision variable that indicates whether a site is selected as a switching center.

5 Consider a transshipment problem with the flows shown in Table 2.2. In this table the ijth entry (ith row, jth column) is the flow from node i to node j.

   a. Draw a diagram showing the nodes and flows.

   b. Identify which nodes are sources, which nodes are destinations, and which nodes are transshipment nodes.

   c. Find the appropriate a_i and b_i for each node.

Table 2.2 TRANSSHIPMENT PROBLEM Flows

	1	2	3	4	5
1	0	0	4	6	0
2	0	0	2	0	4
3	0	0	0	3	3
4	4	0	0	0	1
5	0	2	0	0	0

d. Is the net "product" generated equal to the net product consumed?

6 Answer each part briefly:

a. Describe in words the meaning of the equation

$$\sum_{j=1}^{N} x_{ij} = \sum_{k=1}^{N} x_{ki}$$

for certain i.

b. Let $x_{ij} = 1$ if system model i is installed at site j. There are m models and N sites. What does the following equation *physically* mean?

$$\sum_{i=1}^{m} x_{ij} \geq 1 \qquad j = 1, 2, \ldots, N$$

c. Let x_{ij} be 1 if switch i is connected to switch j and zero otherwise. Let there be N switches. What do these constraints *physically* mean?

$$\sum_{j=1}^{N} x_{ij} \leq 5 \qquad i = 1, 2, \ldots, N$$

$$\sum_{j=1}^{N} x_{ij} \geq 1 \qquad i = 1, 2, \ldots, N$$

7 Write a mathematical program for a modified capital budgeting problem with the goal of minimizing the cost expended to build switching centers while delivering at least a minimum amount, P, of profits. Include any other reasonable constraint and define all variables.

8 Consider a capital budgeting style problem featuring a cellular telephone provider preparing a three-year plan.

At the start of the first year of the plan, a number of franchises are purchased out of N potential franchise locations. Let c_i be the cost of installing infrastructure in the ith franchise location and let x_i be 1 if the ith franchise is purchased (and zero otherwise).

Also let p_{ij} be the predicted profit realized from the ith franchise in year j ($j = 1, 2, 3$).

In the second or third year (but no both) each franchise must be upgraded to handle predicted traffic growth. Let u_{ij} be the upgrade cost at franchise i in year j. Also let y_{ij} be 1 if the upgrade is performed for franchise i in year j ($=2$ or 3). Assume that the upgrade costs can be different between years 2 and 3 but profit, P_{ij}, is not affected by the year in which upgrades occur.

Write a mathematical program to minimize the cost of construction and upgrades while satisfying all relevant constraints. These include certain profit constraints:

a. *Each* franchise makes at least profit P_{site} in year 1.

b. The total system wide (across all franchises) profit is at least $P_{\text{year 1}}$ in year 1.

c. The total system wide profit is at least $P_{\text{year 1-3}}$ for all three years.

9 A number of sites, n, have been selected for the possible construction of satellite ground stations. *One* of m systems is installed at each site. Let

$$x_{ij} = \begin{cases} 1 & \text{if system } i \text{ installed at site } j \\ 0 & \text{otherwise} \end{cases} \qquad (2.128)$$

Let BW_i be the independent bandwidth brought on line by the ith ground station system. Let C_i be the cost of constructing the ith ground station system. Bandwidth and cost do not vary by site but by system. A total of C dollars is available to build the entire network.

Write a mathematical program for this situation that maximizes the total bandwidth brought on line for an investment not greater than C dollars. Show all relevant constraints. Implement in the constraints the availability of only S_i units of the ith system (in stock).

10 A number of sites, n, have been selected for the possible construction of satellite ground stations. At most, *three* of m systems (with not more than one of each) are installed at each site. Let

$$x_{ij} = \begin{cases} 1 & \text{if system } i \text{ installed at site } j \\ 0 & \text{otherwise} \end{cases} \qquad (2.129)$$

Let BW_i be the independent bandwidth brought on line by the installation of the ith ground station system. Let C_i be the cost of constructing the ith ground station system, including installation. Bandwidth and cost do not vary by site but by system.

Write a mathematical program for this situation that minimizes the total cost necessary to bring on line a total bandwidth of at least BW. Show all relevant constraints. Implement in the constraints the availability of only S_i units of the ith system (in stock). Also take into account that the capacity of the air cargo plane that is to deliver the systems in one trip is at most S systems.

11 Consider a U.S. company with separate east and west coast networks that are presently *not* connected. The west coast networks consists of switches (nodes) in Seattle (1), San Francisco (2), and Los Angeles (3). The east coast network consists of switches in New York (4), Washington DC, (5) and Atlanta (6).

A decision is made to connect the two networks through a single transcontinental link that will originate in one of the west coast cities, pass through a hub in either Chicago (A) or Dallas (B), and continue on one of the east coast cities.

Draw a map of this situation. Let c_{ij} be the cost of a link from coast city i (note the ID numbers) to hub city j ($j = A, B$). Let x_{ij} be 1 if there is a link from coast city i ($i = 1, 2, \ldots, 6$) to hub city j ($j = A, B$) and 0 otherwise.

Write a mathematical program that minimizes the cost of the single transcontinental route and satisfies the constraints above.

12 Consider an organization that has bought N ATM switching hubs and installed them. Part of the installation is adding cards to the hubs to handle traffic. There are M different models of cards available.

Let n_{ij} be the *number* of model i cards installed at hub j.

Also, let P_i be the profit associated with the installation of a single model i card. Let C_i be the cost associated with the installation of a single model i card. Do not consider the cost of the hub itself. Cost and profit vary by model, not switch hub.

Write a mathematical program that maximizes profit with an investment no greater than C dollars.

Also, include the following constraints:

a. Each hub has between 1 and 10 cards of all types in *total*.

b. Each hub has no more than one card of model 7.

c. There is a card shortage. Only 12 model 5s are available.

13 Solve the capacity expansion problem on a link described by Tables 2.3 and 2.4 using dynamic programming. In particular, draw and label the installation sequence decision tree.

TAblε 2.3 Predicted Circuit Demand

Year	Predicted Demand (circuits)
1	200
2	500
3	700

Table 2.4 Capacity and Cost

System	Capacity (circuits)	Cost
A	100	$1000
B	200	$1500
C	500	$2200

14. A new national online service expects the demand and cost for its services, as described by Tables 2.5 and 2.6, over the next three years.

Table 2.5 Predicted Demand

Year	Predicted Demand (users)
1	9,000
2	14,000
3	19,000

Table 2.6 Capacity and Cost

System	Capacity (circuits)	Cost
A	5,000	$10,000
B	10,000	15,000
C	15,000	20,000

Create an installation sequence decision tree as in Figure 2.3. Show, as in our diagram, the system(s) to be installed associated with each branch and the cost of each installation sequence.

CHAPTER 3

NETWORK ALGORITHMS FOR PLANNING

3.1 INTRODUCTION

There are a wide variety of algorithms that are useful for network design and planning, other than the ones mentioned in Chapter 2. Now we sample such algorithms from the communications literature.

One group of algorithms was introduced originally for the generation of large-scale network topologies. The problem here is to determine link locations and certain node locations in a way that serves to minimize cost. Among algorithms to be covered are ones for determining concentrator locations, terminal assignments, and minimal weight spanning trees. The source for this material is work by Boorstyn and Frank [Boor].

A somewhat different look at topological design appears in work by Gerla and Kleinrock [Gerl 77]. In studying packet-switched networks, these researchers use queueing delay as a performance measure. Queueing delay increases exponentially as links become loaded. The effect is thus to generate a traffic routing that spreads traffic between each source–destination pair over a number of routes.

A different approach uses a model that captures the effects of economies of scale in networks. It was developed originally by B. Yaged [Yage]. Yaged assumed costs for circuit-switched networks that were a concave function of capacity. Thus, on a per-circuit basis, large capacity systems are more economical than small capacity systems. The result, in terms of topological design, is to produce sparse topologies, with a limited number of high capacity links.

Finally, a most fundamental problem arising in some of the contexts mentioned earlier is that of shortest path routing. Algorithms by Dijkstra and by Ford, Fulkerson, and Floyd are discussed in the conclusion of this chapter.

3.2   THE TERMINAL ASSIGNMENT PROBLEM

In this problem there are a number of generic "terminals." These are devices such as actual terminals, PCs, or workstations at known locations. Each generic terminal has to be interconnected to the outside world through a generic "concentrator." Depending on the type of networking being carried out, this connection is achieved by means of devices such as front-end processors, multiplexer, or ATM switches [Dorf]. Each concentrator can accommodate up to a given maximum number of terminals.

Let's express this problem as a mathematical program. We will define an integer variable, x_{ij}, as follows:

$$x_{ij} = \begin{cases} 1 & \text{terminal } i \text{ connected to concentrator } j \\ 0 & \text{otherwise} \end{cases} \qquad (3.1)$$

Let c_{ij} be the cost of connecting terminal i ($i = 1, 2, \ldots, N$) to concentrator j ($j \in J$, where J concentrator sites have been selected). This could represent the cost of laying fiber and the cost of interface devices. Then the objective function is

$$\min Z = \sum_{i=1}^{N} \sum_{j \in J} c_{ij} x_{ij} \qquad (3.2)$$

There are two constraints. One indicates that each terminal is connected to exactly one concentrator.

$$\sum_{j \in J} x_{ij} = 1 \qquad i = 1, 2, \ldots, N \qquad (3.3)$$

The second constraint indicates that each concentrator can accommodate at most q terminals:

$$\sum_{i=1}^{N} x_{ij} \leq q \qquad j \in J \qquad (3.4)$$

This is, naturally, an integer linear programming problem. But it should look familiar. It is just the transportation problem of Chapter 2 with $a_i = 1$ and $b_j = q$. A general-purpose linear programming algorithm could certainly be used to solve this problem. However, there are special-purpose algorithms available for its solution.

One such optimal algorithm will now be described [Boor]. Let the c_{ij} be placed in a table where the ijth entry indicates the cost of a connection from terminal i to concentrator j:

$$\begin{bmatrix} 10 & 3 & 5 & 7 \\ 2 & 4 & 8 & 12 \\ 2 & 3 & 4 & 5 \\ 8 & 5 & 10 & 12 \\ 3 & 4 & 6 & 7 \\ 8 & 7 & 6 & 12 \end{bmatrix} \tag{3.5}$$

The solution to this problem will be such that one entry in each row will be selected and not more than q entries in each column will be selected. Initially one goes through the table row by row, selecting in each row the entry having the least cost. However, if during the initial pass this means that the column constraint (say, $q = 2$) would be violated, the row in question is skipped. For the foregoing table, this leads to:

$$\begin{bmatrix} 10 & (3) & 5 & 7 \\ (2) & 4 & 8 & 12 \\ (2) & 3 & 4 & 5 \\ 8 & (5) & 10 & 12 \\ 3 & 4 & 6 & 7 \\ 8 & 7 & (6) & 12 \end{bmatrix} \tag{3.6}$$

Here the selected costs are enclosed in parentheses.

The next step in the algorithm is to go through each blank row in turn and select minimal cost elements in each one. Because of the column constraint, this may require changes in some of the rows where minimal cost elements were selected earlier. For instance, suppose that one is to select the "3" in the fifth row. Then the selected entry in either the second or the third row needs to be reassigned. Suppose that a different entry in the second row is selected. Then, as shown below, one of the entries in the second column could be changed to satisfy the column constraint.

$$\begin{bmatrix} 10 & 3 & (5) & 7 \\ 2 & (4) & 8 & 12 \\ (2) & 3 & 4 & 5 \\ 8 & (5) & 10 & 12 \\ (3) & 4 & 6 & 7 \\ 8 & 7 & (6) & 12 \end{bmatrix} \begin{bmatrix} 10 & (3) & 5 & 7 \\ 2 & (4) & 8 & 12 \\ (2) & 3 & 4 & 5 \\ 8 & 5 & (10) & 12 \\ (3) & 4 & 6 & 7 \\ 8 & 7 & (6) & 12 \end{bmatrix} \tag{3.7}$$

Here the total cost of the table at the left is 25, compared to a cost of 28 for the table at the right.

Alternatively, the entry in the third row may be changed. Again, this necessitates a change in an entry in the second column, as follows:

$$
\begin{bmatrix}
10 & 3 & (5) & 7 \\
(2) & 4 & 8 & 12 \\
2 & (3) & 4 & 5 \\
8 & (5) & 10 & 12 \\
(3) & 4 & 6 & 7 \\
8 & 7 & (6) & 12
\end{bmatrix}
\begin{bmatrix}
10 & (3) & 5 & 7 \\
(2) & 4 & 8 & 12 \\
2 & (3) & 4 & 5 \\
8 & 5 & (10) & 12 \\
(3) & 4 & 6 & 7 \\
8 & 7 & (6) & 12
\end{bmatrix}
\tag{3.8}
$$

The table at the left has a cost of 24 and the table at the right has a cost of 27. The actual algorithm would proceed further to be sure that an optimal solution was found. For instance, suppose that 3 is selected in the fifth row. The third-row entry could then be changed to 4. The total, optimal, cost is now 23.

$$
\begin{bmatrix}
10 & (3) & 5 & 7 \\
(2) & 4 & 8 & 12 \\
2 & 3 & (4) & 5 \\
8 & (5) & 10 & 12 \\
(3) & 4 & 6 & 7 \\
8 & 7 & (6) & 12
\end{bmatrix}
\tag{3.9}
$$

Note that concentrator 4 is never utilized. This is because of the expense of connecting terminals to it compared to the cost of connecting terminals to the other concentrators.

3.3   THE CONCENTRATOR LOCATION PROBLEM

In the terminal assignment problem, the locations of the concentrators were fixed and known. In the concentrator location problem one seeks to select the best locations for concentrators from a list of candidates sites, as well make the terminal assignments for m sites and n terminals.

Again, let

$$
x_{ij} = \begin{cases} 1 & \text{terminal } i \text{ connected to concentrator } j \\ 0 & \text{otherwise} \end{cases}
\tag{3.10}
$$

$$
y_j = \begin{cases} 1 & \text{concentrator placed at site } s_j \\ 0 & \text{otherwise} \end{cases}
\tag{3.11}
$$

The objective function then becomes

$$
\min Z = \sum_{i=1}^{n} \sum_{j=1}^{m} c_{ij} x_{ij} + \sum_{j=1}^{m} d_j y_j
\tag{3.12}
$$

where d_j is the cost of placing a concentrator at site s_j. It can be seen that the first term represents the cost of the terminal assignments and the second term represents the cost of the concentrator placement. This latter cost would include the cost of the concentrator itself, as well as the cost of securing a location for it at site s_j.

In addition, two sets of constraint equations are needed to complete this program. The first set of constraint equations expresses a limitation of concentrator capacity, that is,

$$\sum_{i=1}^{n} x_{ij} \leq ky_j \qquad j = 1, 2, \ldots, m \qquad (3.13)$$

where k is the maximum number of terminals that can be connected to one concentrator. Of course, if a concentrator is *not* placed at site s_j, then $y_j = 0$ and no terminals are made assigned to that site.

The second set of constraints forces each terminal to be assigned to one concentrator:

$$\sum_{j=1}^{m} x_{ij} = 1 \qquad i = 1, 2, \ldots, n \qquad (3.14)$$

If reliability were an important consideration and one wished to force each terminal to be connected to q concentrators, one would write

$$\sum_{j=1}^{m} x_{ij} = q \qquad i = 1, 2, \ldots, n \qquad (3.15)$$

What is the natural trade-off in the concentrator location problem? If concentrator costs dominate terminal assignment costs, there are likely to be few concentrators. Each concentrator will be connected to many terminals (Fig. 3.1a). On the other hand, if the cost of concentrators is dominated by terminal assignment costs, there will be many concentrators. Each will be connected to a small number of nearby terminals (Fig. 3.1b).

Those familiar with the field of operations research will recognize that this problem is similar to location problems with plants, warehouses, or facilities [Hand] [Fran 74]. For instance, one could envision a situation involving a supermarket chain that had to place warehouses (concentrators) in a way that minimized warehouse costs and shipping costs from warehouses to supermarkets (read terminal assignment costs). Note that, mathematically, the formulation of both problems is the same.

Problems of this type are integer linear programming problems. However, because of the difficulties of solving such a formulation, exhaustive search (for small networks) and branch and bound algorithms (see Chapter 2) are often used.

It is possible to develop heuristic algorithms [Boor] to solve large instances of the concentrator location problem. Why are these needed? With m possible

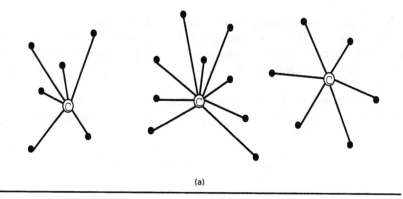

(a)

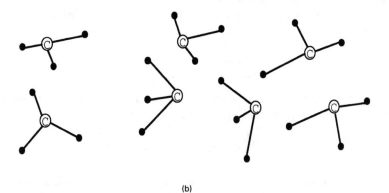

(b)

Figure 3.1 Two possible concentrator locations and terminal assignments.

concentrator sites, there are clearly 2^m patterns of placing concentrators. Even if one restricts attention to placing l concentrators in m sites, the number of patterns of placing concentrators is:

$$\binom{m}{l} = \frac{m!}{l!(m-l)!} \tag{3.16}$$

For instance, for only 30 sites with exactly 15 concentrators there are 155 million patterns of placing the concentrators. Thus even for moderate-sized problems there can be too many patterns for an exhaustive search and possibly too many for a branch and bound algorithm.

Two heuristic approaches, known as ADD [Kueh] and DROP [Feld], can be used as the basis of heuristic algorithms for large-scale versions of the concentrator location problem. Let's consider the ADD algorithm first. One starts with all terminals assigned to a centrally located site. Next one tries to find the best second concentrator to add to the system. That is, for each of $(m - 1)$ possible sites, one places a concentrator there and reassigns up to k (the minimum concentrator capacity) terminals to it from the first site. The k terminals selected should result in the largest savings. The concentrator site selected is the one at which placing the second concentrator results in the minimum cost. Then the terminal assignment problem is solved optimally for this two-concentrator $(1 = 2)$ system, and the cost is recorded.

Next one considers adding a third site. In turn, the terminal assignment problem for the original two concentrators is solved and a concentrator is placed tentatively at the one of the remaining $(m - 2)$ empty sites. Then a concentrator is placed permanently at the site resulting in the minimum cost, and this cost is recorded. In the same way, the minimal cost location is found for a fourth concentrator, a fifth concentrator, and so on, until no improvement in the cost results. This is then the solution.

If one thinks about it, one can see that the terminal assignment problem has to be solved $(m - 1) + (m - 2) + \cdots + (m - p)$ times, where p is the number of concentrators ultimately found. This is a total of mp evaluations. As Boorstyn and Frank point out, if an optimal algorithm evaluated all patterns of concentrator locations consisting of up to p concentrators, there would be a much larger number of evaluations approximately equal to m^p [Boor]. For instance, if there are 30 sites and 7 concentrators, then $mp = 210$ evaluations and $m^p = 22$ billion evaluations!

However, the price for the much smaller number of terminal assignment evaluations is the absence of a guarantee that an optimal solution will be produced. In fact, some concentrators that are sequentially incorporated into the solution really should not be kept in the solution. For instance [Boor], if the first site is in the center of the terminal solution, the optimal solution for two concentrators may in fact include one in the east and one in the west. More sophisticated versions of ADD could periodically and heuristically evaluate potential sites and discard those that have only a minimal impact on the cost function.

In the DROP algorithm one moves in the contrary direction. Initially a concentrator is placed at each site and the terminal assignment problem is solved. Poorly utilized concentrators are dropped until there is no improvement in cost.

3.4   MINIMUM WEIGHT SPANNING TREES

3.4.1   Introduction

A spanning tree is a special type of graph that serves to connect N nodes. Each node is connected to one or more links (edges). The overall graph, being a tree, has no cycles (loops). A little thought will show that a spanning tree connecting N nodes consists of exactly $N - 1$ edges. One can assign "weights" (really costs), c_{ij}, to each possible link. Then the minimal weight spanning tree (MWST) problem is to find a spanning tree that minimizes the sum of the weights of the links chosen. For instance, if the weights represented the distances between the different nodes, one would be calculating the spanning tree of minimum total length. If the weights represented the cost of each link, one would be calculating the minimum cost network.

Why are spanning trees important? Actual networks can, in some cases, be laid out in this manner. Each of $N - 1$ nodes may be a generic "terminal" (i.e., a PC or workstation) and one node may be a central site (computer). Each generic terminal accesses the computer for external communications. Such a network layout is called a multidrop network. A branch of the network hanging off the central site is called a multidrop line. A **polling protocol** [Schw] may be used to provide two-way communication. That is, the central site will send a message to each station on a branch, sequentially, indicating that it may go ahead and transmit and receive messages. In a one-way CATV (cable antenna television) system, on the other hand, the central site may simply be a **head-end** station. The head end originates broadcast signals that travel in one direction, down a tree network to the terminals.

We will consider two versions of the MWST problem in accordance with the survey cited earlier [Boor]. In the unconstrained optimum MWST problem, there is no limit to the number of nodes that can be on a single branch hanging off the central site. Efficient and optimal algorithms are known for this version of the problem. In the constrained MWST spanning tree problem, each branch can have no more than v nodes. This situation might arise because of the limitations of the electrical interfaces used at the central site to accommodate a branch's traffic (i.e. port speed). Optimal algorithms for this version of the problem tend to be slow, so a number of heuristic algorithms are discussed.

3.4.2 Unconstrained Optimum MWST Problem

We shall consider two algorithms for the unconstrained minimum weight spanning tree problem—Kruskal's algorithm and Prim's algorithm.

Kruskal's Algorithm.   Here one orders the distances between every pair of nodes from smallest to largest. One connects the two nodes that are the smallest distance apart with a link. One then goes down the list, connecting nodes that are the next smallest distance apart. There is only one exception: a link is never chosen if a loop would be formed in the graph. It can be proved [Krus] that this algorithm will produce an optimal unconstrained MWST.

There are two computational costs associated with Kruskal's algorithm. One is the cost of ordering the distances between node pairs. The other cost, which is due to checking for loops, can be handled simply by means of a labeling scheme. As Kruskal's algorithm proceeds, clusters of connected nodes are formed. A loop is created if two nodes within the same cluster are connected. Thus nodes in each cluster should be labelled with a cluster identification number. If a link is placed so that it connects two nodes with different cluster identification numbers, one is simply merging two disjoint clusters. The nodes in the new merged cluster receive a new cluster identification number. However, if the two nodes to be connected by the proposed link have the same cluster identification number, a loop would be formed and this link placement is rejected. Node labeling can be performed quite efficiently.

The dominant cost in running Kruskal's algorithm for large networks may well turn out to be the ordering of shortest path distances. With N nodes there are $N(N-2)/2 = \mathcal{O}(N^2)$ node pair distances to consider.[1] Generally, a direct ordering of M numbers requires M^2 comparisons. The use of a heap-type data structure can reduce this to $\mathcal{O}(M \log M)$ or $N^2 \log N$ comparisons.

What is a heap? A heap is a binary tree in which the value associated with each node is, say, less than or equal to the values associated with its two leaves. A heap is particularly easy to store as a column vector H. The value associated with the jth node is $H(j)$. The leaves of the jth node are nodes $2j$ and $2j + 1$. The associated values are stored in $H(2j)$ and $H(2j + 1)$, respectively. The root is node 1 with value $H(1)$. At any time the smallest value is associated with the root. It should be clear that one could, alternatively, set up a heap to find a maximum and one to find a minimum.

Initially, the entries of $H(j)$ can be entered in any order. They can then be arranged to form a heap, using a number of comparisons that is proportional to the number of elements in the heap [Aho]. After each smallest distance is read from the root [$H(1)$] the heap can be reconstituted so that the next smallest element appears in $H(1)$ using $\log_2 M$ comparisons (there are $\log_2 M$ levels in an M-element binary heap).

Actually $N^2 \log N$ comparisons is an upper bound. This is because N^2 numbers are ordered, but far fewer (somewhat bigger than N) are selected. A more careful analysis [Boor] would assume that αN branches are selected, where $1 \leq \alpha \leq N$. Then the number of comparisons is

$$M + \alpha N \log M \sim N^2 + \alpha N \log N^2 \sim \mathcal{O}(N^2) \qquad (3.17)$$

A smaller measure of computational complexity results [Boor] if the network is sparse. The **degree** of a node is the number of links it is connected to. If the average degree of nodes in the network is d, then the total number of branches in the network that must be ordered is dN. The complexity is now:

$$dN + \alpha N \log dN \qquad (3.18)$$

---

[1] Here $\mathcal{O}(N^2)$ indicates the highest polynomial order of complexity. See any book on algorithm theory [Aho] [Sedg].

If all the branches are ordered, there must be $dN \log dN \sim dN \log N$ comparisons.

Prim's Algorithm. The second unconstrained minimum weight spanning tree algorithm is Prim's algorithm. One starts with a node near the center of the network. Its nearest neighbor is connected to it. This yields a component of size 2. Now the node that is closest to that component (either node) is connected. This is found by computing the distances from each node to the first node that was added to the component and comparing these with the distances to the center node. The distances to the center node were computed earlier and are assumed to be stored.

Now a component of size 3 is formed. The process is repeated. In each step one finds the minimum distance between the unattached nodes and the nodes in the component formed so far. More specifically, in each step one looks for the distance between the unattached nodes and (a) the most recently added node in the component (which must be computed) and (b) the other nodes in the component (which have been stored). Naturally, some computational effort can be saved if one stores only the closest node in the component to each unattached node. In the ith of N steps, one must compute the minimum of $(N - i)$ distances for a total of $N(N - 1)/2$ or $\mathcal{O}(N^2)$ comparisons. As is pointed out in by Boorstyn and Frank [Boor], this can be improved if one takes into account the sparsity of the data structure and the utilization of a heap-type data structure.

3.4.3 Heuristic Algorithms for the Constrained Problem

A number of heuristic algorithms have been proposed for the constrained minimum weight spanning tree problem. In this problem each branch connected to the central node can contain no more than v nodes. This feature captures a possible limitation of multidrop lines. Optimal algorithms for this problem can take an amount of solution time that grows as an exponential function of the problem size. Heuristic algorithms can be much more efficient, though the solutions produced generally will be close to optimal, rather than guaranteed to be optimal. Three heuristic algorithms for the constrained minimum weight spanning tree problem are now presented, along with examples from our usual source [Boor].

Modified Kruskal Algorithm. It is assumed that one of the nodes is identified as being the root node. The modified Kruskal algorithm selects nodes in a manner similar to the original Kruskal algorithm. However, in the modified Kruskal algorithm when a cluster of connected nodes contains v nodes, it is connected directly to the root from the closest node in the cluster. The cluster is then assumed to be finished and plays no part in the rest of the algorithm's run.

Naturally, clusters are not merged if their combined size is greater than v nodes. The modified Kruskal algorithm is about as efficient as the original Kruskal algorithm, but its solution is not necessarily optimal. The problem, as related by Boorstyn and Frank, is that the modified algorithm may connect nodes in clusters near the root (center) first. It is then necessary to run long lines to nodes near the periphery.

For instance, consider Figure 3.2. In this diagram the scale is such that the distance between node pairs $(0, a)$, (a, b), and (b, c) is 1. The distance between nodes c and d is 2. Also, in this and the following examples we assume that the maximum number of nodes per cluster is three. With a constraint of $v = 3$, the modified Kruskal algorithm will create the spanning tree in Figure 3.2a. However this is not optimal. The optimal spanning tree appears in Figure 3.2b, where the total cost is 6, compared to a cost of 8 for the first spanning tree.

Esau and Williams Algorithm.   This more sophisticated algorithm utilizes a function of node pair distances, rather than the node distance itself, in performing an ordering of distances. Let c_{ij} be the distance between node i and node j. Let d_i be the distance from the closest node in the ith cluster to the center node. Then form $c'_{ij} = c_{ij} - d_i$. Order the c'_{ij}. The algorithm otherwise proceeds as Kruskal's algorithm does, using the c'_{ij}. However the ordering may change as new clusters are formed and existing clusters grow. Basically, the ordering is now based on the distance from a node to its access node to the center node. Because of the changing nature of the ordering, the most efficient way to implement the Esau and Williams algorithm probably is with a heap-type data structure.

The Esau–Williams algorithm would find the optimal constrained spanning tree in Figure 3.2b. This is not true, however, of the situation that appears in

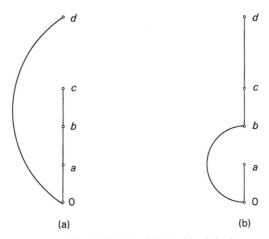

(a)   (b)

Figure 3.2 Modified Kruskal algorithm behavior.

Figure 3.3a, where nodes c and d are somewhat closer to each other than to nodes b and e. The Esau–Williams algorithm would form links (a, b), (c, d), and (e, f). Since merging any of these clusters would violate the multidrop constraint, all three clusters would be connected to the center node. The total cost is thus 3+. By way of contrast, the optimal solution is shown in Figure 3.3b, with a cost of 2+.

As is mentioned elsewhere [Boor], this failure is possible with both the modified Kruskal algorithm and the Esau–Williams algorithm. If two clusters are each of size greater than $v/2$, they cannot be merged. A modified version of the Esau–Williams approach is Whitney's algorithm [Elia] [Whit 72].

Sharma–El Bardai Algorithm.   To understand this algorithm [Shar 70], let there be a ray originating from the center node. As it is rotated, the nodes passed over by the ray are formed into a cluster. When such a cluster contains v nodes, it is connected to the center node via the closest node in the cluster. This algorithm will produce an optimal solution for the graph of Figure 3.3b. However this may not always be true. For instance, as in Boorstyn and Frank, consider the node pattern in Figure 3.4a. The Sharma–El Bardai algorithm will form three clusters: (a, d, g), (b, e, h), and (c, f, i). The total cost is 9. The optimal constrained spanning tree appears in Figure 3.4b, with a cost of 6+. In fact, the modified Kruskal algorithm and the Esau–Williams algorithm will find the optimal solution for this node pattern. All this demonstrates that while these algorithms can produce close to optimal solutions for a wide variety of cases, it is always possible to create node patterns that will lead to very nonoptimal solutions.

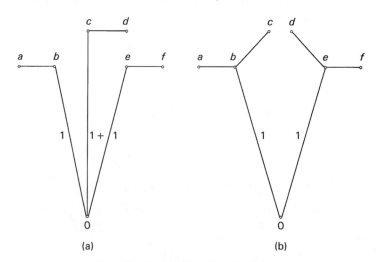

Figure 3.3 Esau–Williams algorithm behavior.

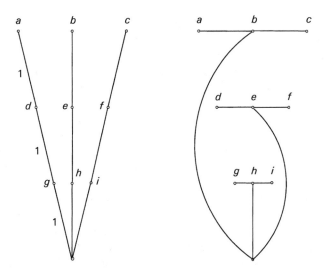

Figure 3.4 Sharma–El Bardai algorithm behavior.

Optimum Algorithms.   Optimal solutions for constrained minimum weight spanning tree problems can be found for small or moderate-size problems using branch and bound algorithms. Upper bounds on the solution for this purpose can be found from any of the preceding algorithms. Lower bounds can be found by means of an unconstrained, minimum weight spanning tree algorithm. Each branch in the decision tree often indicates whether a network branch is included. Then the decision tree has a depth of b (branches) and a width at the bottom (base) of 2^b.

3.5   DISTRIBUTED NETWORK TOPOLOGICAL DESIGN

3.5.1   Introduction

Early computer networks used a centralized architecture. Many terminals would be connected in a star-type pattern to a large mainframe host. Sometimes tree networks were used for this purpose. In both cases a key feature of the design was the absence of a routing problem, since there was only a single path from each terminal to its mainframe.

With the development of packet switching technology in the 1960s, the situation changed. In a packet-switched network digital data is broken into small groups ("packetized") for transmission. Each packet consists of control information such as the destination address (header) and the information payload. People using packet networks began to think of computer networks as suitable for resource sharing [Gerl 77] applied to file transfers, remote logins, and so on. That is, there would be traffic demand beween nodes at the same level in a network hierarchy, instead of simply demand between nodes and a control site. Rather than being centralized, traffic demand becomes distributed.

Moreover, with packet switching it becomes natural for many users to have their asynchronous communications share the same link(s). This asynchronous sharing is sometimes referred to as statistical multiplexing. Because of the bursty nature of data traffic, statistical multiplexing makes for very cost-effective (read low cost) communications. This is especially true in comparison to traditional telephone network circuit switching.

In such distributed networks the tree or star topology no longer makes sense. Instead, the graphs used to represent the network topology have multiple paths between any pair of nodes. There is thus the new dimension of a routing problem. This is even more challenging, as we shall see, because the determination of the optimal routes depends on the capacity assigned to each link and the optimal allocation of capacity to the links depends on the routes chosen. Still, early in the development of computer networks a great deal of progress was made on this problem, which in this section will be recounted.

3.5.2 Queueing Delay as a Penalty

The basic approach to be taken [Gerl 77] is to use the queueing delay on links as a penalty function to drive the optimization. That is, excessive queueing of packets will lead to an excessive delay, which will serve to penalize solutions involving heavy utilization of links. Penalty function methods are discussed in any book on optimization theory.

Let's proceed to model this queueing delay. Queueing models can be expressed either in terms of continuous time models or discrete time models [Robe 94]. In the following, we shall use a continuous time model on the grounds that events in a distributed packet switched network occur asynchronously.

The most basic of continuous time queueing models is the M/M/1 queue. It comprises a single server and single waiting line. It is discussed extensively in texts on queueing theory [Gros] [Klei 75] [Robe 94]. For such a queueing system the arrival process is Poisson (i.e., random), and the service time is a random variable that is negative exponentially distributed. With these assumptions, the queue is memoryless. That is, the past history of the queue has no predictive power. The state of the queue is then simply the current number of customers in the queueing system. This fact about a memoryless system drastically simplifies the associated analysis.

It is well known in queueing theory that the average number of packets is an M/M/1 queue in equilibrium is

$$E(n) = \frac{\rho}{1 - \rho} \tag{3.19}$$

Note that this expression is for an M/M/1 queue with no limit on buffer size. Here $\rho = \lambda/\mu$, where ρ is utilization, λ is the mean arrival rate of packets, and μ is the mean service rate of packets. Since packets must be cleared out of the queueing system at least somewhat faster than they arrive, $\lambda < \mu$ or $\lambda/\mu < 1$. Note from this expression that the expected number of packets increases in an almost exponential-like manner as $\rho \to 1$ (see Figure 3.5).

Another basic queueing relationship is Little's law, which is stated as follows:

$$E(n) = \lambda\tau \tag{3.20}$$

where $E(n)$ is the expected number of packets in the queueing system, λ is the mean arrival rate, and τ is the mean delay a packet experiences in passing through the queuing system. Putting equations (3.19) and (3.20) together, one has:

$$\tau = \frac{E(n)}{\lambda} = \frac{1/\mu}{1 - \lambda/\mu} = \frac{1}{\mu - \lambda} \tag{3.21}$$

Let's try to use this relation as the basis for an expression for the delay on the ith link. Here λ will simply become λ_i. Moreover μ is replaced by μC_i. Now $1/\mu$ is the average packet length in bits per packet and C_i is the channel capacity in bits per second. The latter quantity, channel capacity, is the maximum rate at which

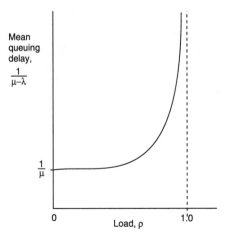

Figure 3.5 Mean queueing delay as a function of load.

information can be transferred through a particular channel. So the delay in the ith link is

$$\tau_i = \frac{1}{\mu C_i - \lambda_i} \tag{3.22}$$

The next step is to construct an expression for the mean delay averaged over all the links. To do this let γ_{jk} be the average traffic rate between source node j and k. Then

$$\gamma = \sum_{j,k} \gamma_{j,k} \tag{3.23}$$

Now define network-wide mean delay as

$$T = \sum_{i=1}^{l} \frac{\lambda_i}{\gamma} \tau_i \tag{3.24}$$

where l is the number of links. Then

$$T = \frac{1}{\gamma} \sum_{i=1}^{l} \frac{\lambda_i}{\mu C_i - \lambda_i} \tag{3.25}$$

$$T = \frac{1}{\gamma} \sum_{i=1}^{l} \frac{\lambda_i/\mu}{C_i - \lambda_i/\mu} \tag{3.26}$$

Let $f_i = \lambda_i/\mu$ be the mean flow rate in bits per second. Then

$$T = \frac{1}{\gamma} \sum_{i=1}^{l} \frac{f_i}{C_i - f_i} \tag{3.27}$$

Note that as f_i approaches C_i on any link, the delay increases radically. Thus one can expect that any optimization technique using this as the objective function will spread the traffic flow so that no link (route) is too heavily utilized.

One might wonder about the possibility of using a queueing model other than the M/M/1 model. In fact, an expression for mean delay on a link can be used from the M/G/1 queueing model. The "G" refers to a model with a general service time distribution. That is, the service times are independent and identically distributed random variables following a known service time distribution, $b(t)$. It has been known since the 1930s that the expected number of customers in such a queueing system obeys the Pollaczek–Khinchin mean value formula:

$$E(n) = \rho + \frac{\rho^2 + \lambda^2 \sigma_s^2}{2(1 - \rho)} \tag{3.28}$$

Here $\rho = \lambda/\mu$, where λ is the mean arrival rate (of a Poisson process) and μ is the mean service rate. Also σ_s^2 is the variance of the service time distribution.

If we replace μ with μC_i as before through some algebraic manipulation, an expression for the mean delay in the ith link can be developed:

$$T_i = \frac{1}{\mu C_i}(1 - \beta) + \frac{\beta}{\mu C_i - \lambda_i} \qquad (3.29)$$

Here $\beta = (1 + \mu^2\sigma^2)/2$. For the special case of an M/D/1 queueing model, the service times are deterministic quantities. Thus, $\sigma_s^2 = 0$ and

$$T_i = \frac{1}{2\mu C_i} + \frac{1}{2(\mu C_i - \lambda_i)} \qquad (3.30)$$

An even more general expression for mean delay is [Klei 70]:

$$T_i = \frac{1}{\gamma}\sum_{i=1}^{l} \lambda_i[T_i + P_i + K_i] \qquad (3.31)$$

The additional delay factors of propagation delay on the ith link (P_i) and nodal interface processing delay (K_i) are assumed to be negligible in the discussion that follows.

3.5.3 The Capacity Assignment Problem

Introduction. In this limited problem it is assumed that the network topology is known. There is a traffic matrix **R** where the ijth entry, r_{ij}, is the predicted mean traffic flow from node i to node j. In reality traffic statistics are hard to predict exactly and many fluctuate widely. It is assumed that the routing policy is known so that, as a consequence, the flow vector is known:

$$\mathbf{f} = (f_1, f_2, \ldots, f_l) \qquad (3.32)$$

where f_i is the mean flow on the ith link. The objective function is to minimize the network mean cost

$$D = \sum_{i=1}^{l} d_i(C_i) \qquad (3.33)$$

where the cost of the ith link is a function d_i of the link capacity, C_i. The number of links is l.

In this capacity assignment problem the objective is to allocate capacity to each link so at to minimize cost. The capacity vector is

$$\mathbf{C} = (C_1, C_2, \ldots, C_l) \qquad (3.34)$$

where C_i is the capacity on the ith link. There are two substantive constraints. One is

$$\mathbf{f} \leq \mathbf{C} \qquad (3.35)$$

This constraint states that, on an element-by-element basis, mean flow is less than capacity. The second constraint holds that mean network delay must be less than some upper limit, T_{max}:

$$T = \frac{1}{\gamma} \sum_{i=1}^{l} \frac{f_i}{C_i - f_i} \leq T_{max} \qquad (3.36)$$

Taken together, the foregoing elements can be expressed as the mathematical program [Gerl 77] outlined in Table 3.1.

The capacity assignment problem will now be examined under a variety of cost functions.

Linear Cost with Fixed Charge.   Let

$$d_i(C_i) = d_i C_i + d_{i0} \qquad (3.37)$$

The costs are proportional to capacity, and there is a fixed charge, d_{i0}, which is independent of capacity. The optimal capacity, allocation can be obtained using the method of Lagrange multipliers [Brys] [Klei 64]. The optimal capacity to be assigned to link i is

$$C_i = f_i + \frac{\sum_{j=1}^{l} \sqrt{d_j f_j}}{\gamma T_{max}} \sqrt{\frac{f_i}{d_i}} \qquad (3.38)$$

The minimum cost is then

$$D = \sum_{i=1}^{l} \left[d_i f_i + d_{i0} + \frac{(\sum_{j=1}^{l} \sqrt{d_j f_j})^2}{\gamma T_{max}} \right] \qquad (3.39)$$

Concave Costs.   One possibility is to linearize the cost and solve a series of linear models in an iterative manner. There will likely be very many local minima in this case [Gerl 73]. Kleinrock [Klei 70] considered the case in which the cost

Table 3.1 Program for Dealing with a Capacity Assignment Problem

Given	Topology
	Traffic matrix **R**
	Routing policy [i.e., $\mathbf{f} = (f_1, f_2, \ldots, f_l)$]
	Cost-capacity function $d_i(C_i)$
Minimize	$D = \sum_{i=1}^{l} d_i(C_i)$
Over	$\mathbf{C} = (C_1, C_2, \ldots, C_l)$
With constraints	$\mathbf{f} \leq \mathbf{C}$
	$T = \frac{1}{\gamma} \sum_{i=1}^{l} \frac{f_i}{C_i - f_i} \leq T_{max}$

function follows a power law:

$$d_i(C_i) = d_i C_i^\alpha + d_{i0} \qquad 0 \le \alpha \le 1 \tag{3.40}$$

In this case there is a unique local minimum.

Discrete Costs. The capacity assignment problem can be solved either optimally by dynamic programming [Gerl 73] or suboptimally by Lagrangian decomposition [Ever] [Fox]. The dynamic programming algorithm of Gerla has a computational cost that is quadratic in the number of links. The cost of Lagrangian decomposition is somewhat more than linear. Naturally this approach is suboptimal.

3.5.4 A Routing Problem

In this limited problem the channel capacities are known and the objective is to find the optimal routes, and thus flows, for the traffic. The problem can be formulated according to Table 3.2 [Gerl 77].

Note here that in the objective function $P_i + K_i$ is multiplied by $\mu f_i = \lambda_i$ from equation (3.26). This problem is in fact a convex multicommodity flow problem on a convex constraint set.

There are a variety of optimal solution methods for the multicommodity flow problem [Dant]. However, they tend not to be very efficient. Thus a number of suboptimal techniques are possible [Gerl 77]. However the remainder of this section focuses on a downhill search algorithm called the flow deviation algorithm, which can find the optimal solution for this routing problem. Moreover, as a heuristic algorithm, the flow deviation is computationally efficient.

The following properties of the optimal solution will prove useful in developing the flow deviation algorithm.

Property A [Gerl 77] It turns out that the set of multicommodity flows **f** that satisfy the traffic matrix **R** is a convex polyhedron, as in linear programming. The corners of the polyhedron have a special significance. Each "extremal flow" corresponds to a shortest path routing under some assignment of weights to links.

Table 3.2 Program for Dealing with a Routing Problem

Given	Topology
	Channel capacities **C**
	Traffic matrix **R**
Minimize	$T = \dfrac{1}{\gamma}\sum_{i=1}^{l} f_i\left[\dfrac{1}{C_i - f_i} + \mu(P_i + K_i)\right]$
Over	$\mathbf{f} = (f_1, f_2, \ldots, f_l)$
With constraints	$\mathbf{f} \le \mathbf{C}$
	f is a multicommodity flow satisfying **R**

Naturally, any flow in the convex polyhedron can be expressed as a convex combination of the extremal flows.

Property B [Gerl 77] In the flow deviation algorithm, for any multicommodity flow **f**, a link weight for each link can be defined as a function of mean network delay T. Specifically, $w_i = \partial T / \partial f_i$.

As in Gerla and Kleinrock [Gerl 77], let ϕ be the shortest path routing associated with these link weights. Consider now:

$$\mathbf{f}^{(j+1)} = (1 - \alpha)\phi + \alpha \mathbf{f}^{(j)} \qquad (3.41)$$

where α is chosen to minimize time delay. Also, $\mathbf{f}^{(j)}$ is a set of flows at the jth iteration.

The flow deviation algorithm now proceeds as follows:

Step 0 Let $\mathbf{f}^{(0)}$ be the initial feasible flow.

Step 1 Compute ϕ^j, the shortest path routing (see Section 3.7) such that $w_i^{(j)} = [\partial T / \partial f_i]_{\mathbf{f} = \mathbf{f}^{(j)}}$ for each $i = 1, 2, \ldots, l$.

Step 2 Let

$$\mathbf{f}^{(j+1)} = (1 - \alpha_j)\phi^{(j)} + \alpha_j \mathbf{f}^{(j)} \qquad (3.42)$$

Here α_j $(0 \le \alpha_j \le 1)$ is selected to minimize mean network time delay:

$$\min_{\alpha} T[(1 - \alpha)\phi^{(j)} + \alpha \mathbf{f}^{(j)}] \qquad (3.43)$$

Step 3 If

$$|T(\mathbf{f}^{(j+1)}) - T(\mathbf{f}^{(j)})| < \varepsilon \qquad (3.44)$$

then stop. This is the termination criteria. Otherwise let $j = j + 1$ and go to step 1.

As Gerla points out, step 1 is the most extensive step computationally. It takes between n^2 and $n^2 \log n$ operations, where n is the number of nodes.

3.5.5 Combined Capacity and Flow Assignment

The combined problem of capacity and flow assignment requires the simultaneous determination of the optimal routing and captivity assignment as indicated in Table 3.3. For this problem there are very many local minima. As in the case in most optimization problems, this makes finding an optimal solution difficult. Thus only suboptimal algorithms for special cases are discussed below.

Linear Costs. From equation (3.38) it is clear that, for the linear cost case, one can find the optimal capacities for a given flow variable. Therefore in the preceding mathematical program one can, through the use of this expression,

Table 3.3 Program for Dealing with Combined Capacity and Flow Assignment Problem

Given	Topology
	Traffic matrix $\mathbf{R}$
	Cost-capacity functions $d_i(C_i)$
Minimize	$D(\mathbf{C}) = \sum_{i=1}^{l} d_i(C_i)$
Over	$\mathbf{f}, \mathbf{C}$
Such that	$\mathbf{f}$ is a multicommodity flow satisfying the traffic matrix $\mathbf{R}$
	$\mathbf{f} \le \mathbf{C}$
	$T(\mathbf{f}, \mathbf{C}) = \dfrac{1}{\gamma} \sum_{i=1}^{l} \dfrac{f_i}{C_i - f_i} \le T_{\max}$

eliminate the capacities as decision variables. The objective function then becomes

$$D(\mathbf{f}) = \sum_{i=1}^{l} \left[d_i f_i + d_{i0} + \frac{(\sum_{j=1}^{l} \sqrt{d_j f_j})^2}{\gamma T_{\max}} \right] \qquad (3.45)$$

Here the optimization is solely over the decision variable $\mathbf{f}$. However $D(\mathbf{f})$ is now concave, not convex, over the convex polyhedron of multicommodity flows. There are thus many local minima. Each local minimum corresponds to an external flow (see Figure 3.6).

The flow deviation algorithm can be modified to search for local minima [Gerl 77]:

Step 0 Here $\mathbf{f}^{(0)}$ is a feasible initial starting flow.

Step 1 Let $\mathbf{f}^{(j+1)}$ be the external flow at the jth iteration associated with:

$$w_i^{(j)} = \left[\frac{\partial D}{\partial f_i} \right]_{\mathbf{f}=\mathbf{f}^{(j)}} \qquad \forall \, i = 1, 2, \dots, l \qquad (3.46)$$

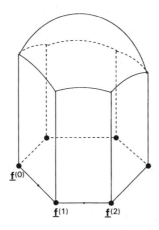

Figure 3.6 Feasible solution polyhedron of combined problem.

Step 2 If $D(\mathbf{f}^{(j-1)}) \geq D(\mathbf{f}^{(j)})$, stop. Then $\mathbf{f}^{(j)}$ is a local minimum. Otherwise set $j = j + 1$ and proceed to step 1.

Basically the algorithm evaluates many extremal flows, keeping the best (minimal cost) one.

By differentiating D, as in step 1, one has for the weight for the ith link at the jth iteration.

$$
w_i^{(j)} = d_i \left[1 + \frac{\sum_{i=1}^{l} \sqrt{d_j f_j}}{\gamma T_{\max}} \frac{1}{\sqrt{d_i f_i}} \right]
\tag{3.47}
$$

Which local minimum the algorithm finds depends on the choice of initial flow. To increase the chances of reaching a good solution, it is advisable to try mutiple initial flows.

One interesting property of equation (3.47) is that $\lim_{f_i \to 0} l_i = \infty$. This means that whenever a link flow goes to zero, the link weight then goes to infinity so that the link is no longer considered.

Concave Costs.   It was shown by Gerla [Gerl 73] that $D(\mathbf{f})$ is concave over $\mathbf{f}$ and that local minima can be found by means of a modified version of the flow deviation algorithm.

Discrete Costs.   Two approaches are possible [Gerl 77]:

Approach 1 Iterate between a routing problem with fixed capacities and a capacity assignment problem with fixed routes. The optimization procedure terminates when a local minimum is found.

Approach 2 Use continuous concave costs to interpolate discrete costs. Solve the combined (concave) capacity and flow assignment problem. Then transform the continuous capacities to the smaller (feasible) discrete values. Run the routing problem, based on these capacities, again.

A number of algorithms incorporating these approaches are available [Gerl 77].

3.5.6 The Branch X-Change Algorithm

The branch x-change algorithm [Gerl 77] is a generic algorithm in which one link at a time is repetitively added and one is deleted. The operation of adding a link and deleting a link is also called a local transformation. In a typical branch x-change algorithm the steps would be as follows.

Step 0 Create initial topology.

Step 1 A new link is selected to be added to the network and an existing link is selected to be deleted. The modified network must be at least two connected.

Step 2 An algorithm is used to select new link capacities and flows. The algorithm chosen could be, for example, the combined capacity and flow algorithm of Section 3.5.5. The change made in step 1 is kept if there is an improvement in cost and/or throughput.

Step 3 Stop when all local transformations have been evaluated. Otherwise, proceed to step 1. This method will, naturally, find local minima, but it is not guaranteed to find a globally optimal minimum.

Note that nothing has been said about how the links are selected for addition and deletion. This could be done randomly. A more intelligent choice is discussed in the next section.

3.5.7 The Cut Saturation Algorithm

The branch x-change algorithm calls for an exhaustive search of all local exchanges. In addition, the algorithm tends to be time-consuming once a network has reached a certain size. However, it is possible to speed the computation by more intelligently choosing the links to add and delete. This is done in the cut saturation algorithm.

A **cut** is a group of links whose deletion will result in disconnection of the network into two or more pieces. In a **saturated cut** the traffic on each of the links in the cut has reached the respective link's capacity. As the load, hence the traffic, increases, one of the network's many potential cuts eventually will approach saturation. At this point, for the network to be able to handle more traffic, additional capacity (or additional links) must be added to the saturated cut. The smallest set of links with high utilization that will disconnect the network is called the **saturated cutset**.

The cut saturation algorithm consists of six steps [Boor]:

Step 1, Routing Boorstyn and Frank report using a modified version of the flow deviation algorithm for the routing. Conceivably, other appropriate routing algorithms could be used.

Step 2, Saturated cutset determination This step is performed at each step in the routing.

Step 3, Add-only Determining the saturated cutset gives one a good idea of where an additional link will boost network capacity. However, one must be even more specific. One idea would be to add the shortest possible link across the cutset. However, this choice has been found to change the cutset by a small amount, providing only minimal throughput improvement [Boor]. Another potential choice would be to add a link, to connect the traffic centers on either side of the cut. However, such long links tend to be costly. Therefore one may compromise. Heuristic algorithms can be developed [Boor] that place the added link ends at nodes a moderate distance (say, 2 link hops) from the cut.

Step 4, Delete-only In this step an expensive and lightly utilized link is removed. Let D_i be the ith link's cost, C_i the ith link's capacity, and f_i the actual mean flow on the ith link. Then a possible criterion for expense and underutilization is [Boor]:

$$E_i = \frac{D_i \times (C_i - f_i)}{C_i} \tag{3.48}$$

The link with the highest product of cost and excess capacity excess capacity is eliminated. Some networks are constrained to be at least two connected. Note that if removal of a link would cause a loss of 2-connectivity in such a network the link is not removed.

Step 5, Perturbations When a network with a desired average throughput has been created, one may wish to improve the design. That is, one would like to maintain the throughput while decreasing cost. One possibility for doing this is the branch x-change algorithm. An algorithm with more intelligence would be the use of add-only and delete-only in a combined "perturbation" operation.

Lower and upper bounds can be set up to contain nominal throughput exchanges as a result of perturbation operations. Delete-only can be used solely to bring the mean throughput back from excursions above the upper bound. Add-onlys can be used to bring the mean throughput back from excursions below the lower bound.

Step 6, Chain collapsing The cut saturation algorithm is more efficient if chains of serial nodes are collapsed into an equivalent link. This approach is most successful if the traffic through the chain is "transit" traffic from external nodes, as opposed to "internal" traffic between nodes in a chain [Boor].

3.6 ROUTING WITH ECONOMIES OF SCALE

3.6.1 Introduction

An important aspect of large-scale tecommunications networks is referred to as **economies of scale**; that is, the advantages obtainable because a larger system can serve many users is more cost effective than a smaller system serving fewer users. Of course the larger system will cost more than the smaller system in total dollars. However, if one looks at the total cost divided by the number of users, or cost per user, the larger system may be less expensive. This cost per user is often referred to as the **marginal cost**.

In particular, this section will consider long haul transmission systems, following closely the research on the subject by B. Yaged, whose notation we will use [Yage].

Larger capacity systems can be more economical than smaller capacity systems because of the economies inherent in large-scale operations such as manufacturing and construction. A graphical representation of such costs appear in Figure 3.7.

In Figure 3.7a, total cost is plotted versus capacity. Note that this is a characteristic concave function. Let's say that if the flow in link m is y_m, the total cost of link m is $f_m(y_m)$. Figure 3.7b shows the cost per user (read cost/circuit for a telephone transmission system) versus capacity. It is simply the first derivative of $f_m(y_m)$ with respect to y_m or $f_m'(y_m)$.

Now let there be N nodes and l links. Let **R** be the traffic matrix. That is, r_{ij} is the mean traffic flow between nodes i and j.

$$\mathbf{R} = \begin{bmatrix} r_{11} & r_{12} & \cdots & r_{1N} \\ r_{21} & r_{22} & \cdots & r_{2N} \\ \vdots & \vdots & \ddots & \vdots \\ r_{N1} & r_{N2} & \cdots & r_{NN} \end{bmatrix} \tag{3.49}$$

The general problem examined in this section involves the following objective function:

$$Z = \sum_{m=1}^{l} f_m(y_m) \tag{3.50}$$

That is, the sum of the concave costs of flows on each link is considered. The following conditions are assumed to hold [Yage]:

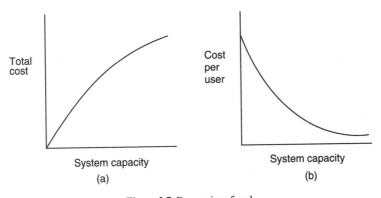

Figure 3.7 Economics of scale.

1 $f_m(y) \geq 0$

2 $f_m'(y) > 0$

3 $f_m''(y) < 0$

4 $f_m(y)$ is continuous

The third condition ensures that the second derivative of f must be less than zero. Thus the function is concave.

When link costs are linear functions of the flow, the following special case of this problem occurs:

$$Z = \sum_{m=1}^{l} f_m(y_m) = \sum_{m=1}^{l} k_m y_m \qquad (3.51)$$

In this case a optimal solution can be found by running a shortest path algorithm (see Section 3.7) on the network graph, where k_m is the weight of the mth link.

3.6.2 Characterizing Local Optimality

The general problem is to determine the optimal routes in a network with concave link costs. This is a minimum cost, multicommodity flow problem with linear constraints, but no link capacity constraints. However, because of the concave cost function there are many possible locally optimal solutions. Therefore it makes sense to first examine the nature of these local solutions. We shall discuss a number of properties of these local solutions from [Yage].

Let an asterisk (*) indicate an optimal routing.

PROPERTY 1

The optimal flow between a pair of nodes follows one and only one path, p_{ij}^*.

THEOREM 1

Property 1 is satisfied by all pairs of nodes in an optimal routing.

Proof

Assume that the flow between nodes i and j takes two or more routes. Then it is straightforward to demonstrate that the cost of routing this traffic can be lowered by transferring it to the route with the lowest total marginal cost. Intuitively, the combination of flow from two or more routes onto a single route is seen to lower the marginal cost. This is a direct result of the cost function of Figure 3.7. ∎

PROPERTY 2

Let traffic flow r_{ij} be optimally routed on path p_{ij}^*. Let nodes a and b be on the path from node i to node j. Then the traffic flow r_{ab} follows the subpath of p_{ij}^* connecting nodes a and b.

THEOREM 2

Property 2 is satisfied for all pairs of nodes by an optimal routing.

Proof

One can use Theorem 1 to show that the entire flow between nodes a and b, be it from node a to b or from node i to j, should be concentrated on a single path from between node a and node b. ∎

THEOREM 3

The collection (i.e., union) of links belonging to the optimal paths between node i and all other nodes j, and for which there is nonzero flow $r_{ij} > 0$, is a tree.

Proof

From Theorem 2 there can be no loops (cycles) in the routing. Moreover, the collection of links is connected, as there is a path from each node to common node i. Thus the collection of links forms a tree. ∎

Let's introduce a new concept called ε-optimality [Yage]. An ε-optimal routing has the following properties:

- ε is an arbitrary small positive number.
- Properties 1 and 2 are satisfied.
- If one attempts to add ε additional units of flow between nodes i and j the optimal path for this flow is the same path taken by the nominal flow between nodes i and j of r_{ij} units.

THEOREM 4

An optimal routing is ε-optimal.

Proof

We will follow closely a proof by Yaged [Yage]. From Theorems 1 and 2 it is known that an optimal routing satisfies Propeties 1 and 2. It will thus be sufficient to show that the presence of Property 1 and 2 implies ε-optimality. A proof by contradiction will be given.

Suppose that an optimal routing satisfies Properties 1 and 2 but is not ε-optimal. Suppose also that an amount of flow, ε, is removed from path P_1 that carries r_{ij} units of flow. The cost of routing the flow r_{ij} on path P_1 is thus reduced by an amount $c_1\varepsilon$, where c_1 is a linear cost constant. It is assumed that ε is small enough that the linear cost model with $c_1\varepsilon$ is valid. Because of the lack of ε-optimality, there is another path, P_2, between nodes i and j to which flow can be added at minimal cost $c_2\varepsilon(c_2 < c_1)$. Thus the total network cost has been reduced by an amount $(c_1 - c_2)\varepsilon$. Thus the original routing was not optimal. The showing of this contradiction proves that if the routing is optimal, it must also be ε-optimal.   ∎

THEOREM 5

For an ε-optimal routing the path for each r_{ij} is the shortest path route between nodes i and j where each link length is $f'_m(y_m)$.

Proof

First, the cost of routing an additional flow of amount ε on link m along a path from i to j is:

$$\varepsilon[f'_m(y_m)] \tag{3.52}$$

Again, it is assumed that ε is small enough to permit the foregoing linear cost model to hold. Note that one has units of flow multiplied by units of cost per flow to yield units of cost.

Now the cost of routing an additional flow of amount ε on link m along a path k, between nodes i and j, is

$$\varepsilon\left(\sum_{m\in p^*_{ij}} f'_m(y_m)\right) \tag{3.53}$$

Thus the minimum cost path can be found using a shortest path algorithm where the weight of each link is $f'_m(y_m)$.   ∎

To state the next theorem the following definitions will be needed:

$$\mathbf{Y} \equiv \text{vector of link flows } \{y_1, y_2, \ldots, y_l\}$$
$$\mathbf{C} \equiv \text{vector of link lengths } \{c_1, c_2, \ldots, c_l\}$$

Here a boldface symbol is a vector. The link flows are a function, A, of the link lengths:

$$\mathbf{Y} = A(\mathbf{C}) \tag{3.54}$$

Also, the link lengths are a function, B, of the flows

$$\mathbf{C} = B(\mathbf{Y}) \tag{3.55}$$

These can be combined in the natural way to form

$$\mathbf{C} = B(A(\mathbf{C})) \tag{3.56}$$

The solution of this equation is C_0. Then

$$\mathbf{C}_0 = B(A(\mathbf{C}_0)) \tag{3.57}$$

Here A is the mapping of lengths to link flows and B is the mapping from link flows to lengths. This is what is referred to as fixed point equation. Here $\mathbf{C}_0$ is the fixed point solution. It is also an implicit equation in that the variables to be found, $\mathbf{C}_0$, appear on both sides of the equation. This cannot usually be solved for in closed form (brought to one side of the equation). Fixed point equations have a long history in science and mathematics.

At this point the nature of the solution of this fixed point theorem can be characterized.

THEOREM 6
A routing that is a solution of the fixed point $\mathbf{C}_0$ of equation (3.57) is ε-optimal.

PROOF
All the properties of ε-optimality are satisfied by the fixed point solution. ∎

Together, Theorems 5 and 6 yield [Yage]:

COROLLARY
Every ε-optimal routing is associated with some fixed point $\mathbf{C}_0$. Moreover, every fixed point $\mathbf{C}_0$ is association with an ε-optimal routing. ∎

Thus the basic iterative algorithm that can be used is

1 Initialize link capacities, $\mathbf{C}_0$.

2 $\mathbf{C}_j = B(A(\mathbf{C}_{j-1})) = \phi(\mathbf{C}_{j-1})$ for $j = 1, 2, \ldots$.

3 Stop when convergence occurs.

Figure 3.8 illustrates the convergence process. Here link flow is plotted against generic link cost. The steplike curve is the demand, $d(\mathbf{C})$. As the steplike nature suggests, for small variations in link cost there may be no change in link

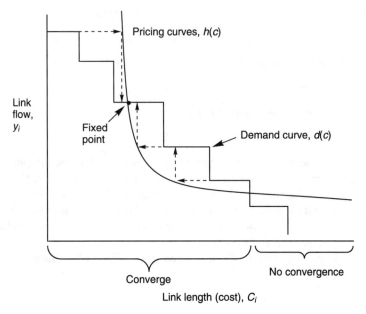

Figure 3.8 Convergence process.

flow. This is because for small variations in link costs, the overall routing may not change.

Also displayed in Figure 3.8 is a "pricing" curve. If $c = f(y)$ is the marginal cost function, then the inverse function, $y = h(c)$, is the pricing curve. The intersection of the pricing curve and the demand curve is the fixed point. If costs are excessive, the link is deleted (see lower right of the figure).

Running the preceding algorithm causes the link flow and cost to converge to a fixed point if the initial starting point is in the convergence region. The algorithm iterates beween link flow allocations based on routing and routings based on link flow allocations.

Because of the concave cost function, the final network synthesized will consist of a minimal number of maximal capacity links and associated routing. The routing is such that traffic between a pair of nodes flows along a single route. This is completely different from the situation in Section 3.5, where a *convex* cost (i.e., delay) function leads to a multiplicity of traffic routes between each pair of nodes. The difference, naturally, is that with a convex cost function the heavy loading of paths is penalized, while with a concave cost function the heavy loading of routes is efficient.

3.7 SHORTEST PATH ROUTING

3.7.1 Introduction

Shortest path routing is the simplest and most commonly used type of routing. The reader may have noticed that such routings are called for in several of the algorithms in this chapter. The basic idea behind a shortest path routing is to represent the network of interest as a graph. The nodes of the graph represent computers, hosts, or switches. The edges of the graph represent communication links. Associated with each edge is a scalar cost value. Typically this may represent physical distance (hence the term "shortest path"), delay, or some security, quality of service, or performance metric.

In a shortest path problem one attempts to find the optimal (shortest) route between one or more given pairs of nodes. Here an optimal path will have a total cost (calculated by summing the costs along the path links) that is minimal. Naturally a shortest path is not necessarily unique.

Routing is usually considered to be part of the network layer in a protocol stack [Tane]. In fact some types of networks such as individual ethernet or token ring local area networks, do not have a routing problem per se because there is a single path between any pair of nodes. However for wide area or interconnected packet-switched or circuit-switched networks, the routing graph is an excellent model.

Routing can be implemented either in a centralized fashion (on a single computer) or in a distributed fashion (on a network of cooperating computers). Most routing algorithms in a single run generate the shortest paths between a given node and all other nodes. It should be clear that this collection of paths is a tree. This is done as easily as generating the shortest path between a single pair of nodes.

In networks with unidirectional links, one can distinguish between two types of algorithm [Saad]. One creates the shortest *forward* path tree from a node to all other nodes in the nework (i.e., Dijkstra's algorithm). The second type of algorithm creates the shortest *backward* path tree, which consists of the shortest paths from all sources to a single destination node (Ford, Fulkerson, and Floyd algorithm). Both algorithms are explained below, though for graphs with bidirectional links [Schw].

3.7.2 Dijkstra's Algorithm

Let's say we are going to find all the shortest paths in Figure 3.9 from node 1 to every other node. The algorithm can be used to create Table 3.4.

There is one column for each node (2, 3, . . . , 6), and the algorithm generates a row at a time. In the table an entry in the (i, j)th position indicates the current

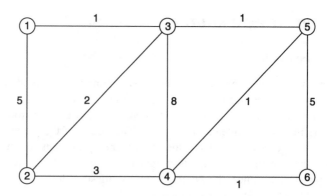

Figure 3.9 Example network for shortest path routing.

distance found from the *j*th node in the *i*th step. Naturally, an actual implementation will store pointers for the current shortest route at each entry.

The initialization phase of the algorithm will fill in the first row of the table. The direct distance between the root node (node 1) and adjacent nodes (nodes 2 and 3) are entered. The entries of nonadjacent nodes are set to a very large value (essentially infinity).

The rest of the table is generated in the following manner. The set *N*, at each row, contains the nodes for which optimal paths to the root have been found so far. For each row we find a node (not in *N*) that has the minimum cost path to the root of all the one-hop neighbors of all the nodes already in set *N*. This node is added to *N*. Such nodes are indicated in parentheses in the table.

The basic recursion that is used to fill in the other entries in that row can be expressed, using the notation in [Schw] as

$$D(v) = \min_{w}[D(v), D(w) + l(w, v)] \tag{3.58}$$

where *w* is the node just added to set *N*. Also *D(v)* is the current distance from node *v* to the root, and *l(w, v)* is the distance along the link connecting *adjacent* nodes *w* and *v*.

Table 3.4 Dijkstra's Algorithm

	N	*D(2)*	*D(3)*	*D(4)*	*D(5)*	*D(6)*
1	{1}	5	1	∞	∞	∞
2	{1, 3}	3	(1)	9	2	∞
3	{1, 3, 5}	3	1	3	(2)	7
4	{1, 3, 4, 5}	3	1	(3)	2	4
5	{1, 2, 3, 4, 5}	(3)	1	3	2	4
6	{1, 2, 3, 4, 5, 6}	3	1	3	2	(4)

Iteration proceeds until an optimal path has been found for each node. Note that all optimal paths may be found before the algorithm terminates, but the algorithm must be run to completion.

One of the first computer network applications of the Dijkstra algorithm was in Arpanet, where it was used to do routing in an environment that had for each node a global database.

As has been mentioned, the set of optimal nodes forms a spanning tree (of $N - 1$ links for an N-node network). In a packet or circuit switching network, the output of a shortest path algorithm may be used to generate a "routing table" for each node. At a node for each incoming packet of circuit the routing table indicates the output link to take to get to a particular destination. For instance, in the example of this section the routing table for node 3 is shown in Table 3.5. For instance, a circuit arriving at node 3 destined for node 4 will be routed first to node 5.

3.7.3 Ford–Fulkerson–Floyd Algorithm

The Ford–Fulkerson–Floyd algorithm also generates a table (Table 3.6). The ijth entry (ith row, jth column) consists of two numbers. The first is the next node along the current path to the root and the second number is the associated total distance. In the initialization each entry is set to $(\cdot, \infty)$.

Table 3.6 is filled in from left to right, top to bottom. The basic recursion for distance implemented for each entry is [Schw]:

$$D(v) = \min_{w}[D(w) + l(w, v)] \tag{3.59}$$

Table 3.5 Routing Table for Node 3

Destination	Output Link
1	to 1
2	to 2
3	—
4	to 5
5	to 5
6	to 5

Table 3.6 Ford–Fulkerson–Floyd Algorithm

	Node 2	Node 3	Node 4	Node 5	Node 6
Initialization	$(\cdot, \infty)$	$(\cdot, \infty)$	$(\cdot, \infty)$	$(\cdot, \infty)$	$(\cdot, \infty)$
1	(1,5)	(1,1)	(2,8)	(3,2)	(5,7)
2	(3,3)	(1,1)	(5,3)	(3,2)	(4,4)
3	(3,3)	(1,1)	(5,3)	(3,2)	(4,4)

Again, $D(v)$ is the current distance from the vth node to the root and $l(w, v)$ is the link distance between neighbors w and v. The minimization is carried out over all the neighbors w of node v.

The basic idea is that at each iteration (row), a node routes through the neighbor with the best (shortest) path. In fact, as one proceeds down the table, information on good routes is shared between neighbors, neighbor's neighbors, and so on. Not surprisingly, the Ford–Fulkerson–Floyd algorithm lends itself to a decentralized implementation.

Let's consider the mechanics of our centralized implementation of the Ford–Fulkerson–Floyd algorithm. To update an entry, one makes use of the most up-to-date information in Table 3.6. This information appears to the left of the entry in the same row and in the row above, to the right of the entry's position. Thus, for instance, when filling in the entry for the second row, fourth node, we can take advantage of a path for neighboring node 5 in the previous row.

Note that the number of rows found in the algorithm's table depends strongly on the labeling of the columns. If one can label the columns starting with the nodes nearest the root (in terms of cost) and proceeding to those furthest from the root, the algorithm will terminate most quickly.

Decentralized implementations of this and other algorithms are sometimes accompanied by additional problems such as looping. That is, as a decentralized algorithm continually updates its routing tables using finite speed links with propagation delays, some packets may "loop" through the network.

Another problem, and potential cause of looping, is that the use of a performance metric such as average measured delay may affect the algorithm in a way that causes (possibly unstable) traffic fluctuations between the parts of the network that are lightly loaded and those that are heavily loaded. This occurs because of the tendency to route traffic through less congested parts of a network, making them congested, and to route traffic away from congested parts of a network, making them less congested. These issues are discussed in detail elsewhere [Bert 87] [Schw].

3.8   PROBLEMS

1 Consider this modified version of the terminal assignment problem. Terminals are either 10,000 or 20,000 bps models.

$$b_i = \begin{cases} 10,000 \text{ bps} & \text{if terminal } i \text{ is a } 10,000 \text{ bps model} \\ 20,000 \text{ bps} & \text{if terminal } i \text{ is a } 20,000 \text{ bps model} \end{cases}$$

Each concentrator can handle a maximum total input from the terminals of 100,000 bps. The cost of connecting terminal i to concentrator j is C_{ij} and if there is such a connection, X_{ij} is one. If there is no such connection, then X_{ij} is zero. There is also an additional connection charge of $100 for each 10,000 bps model and $200 for each 20,000 bps model.

Formulate this problem as a mathematical program. Describe the intention of each equation. How would you handle costs that are not simple multiples of the amount of bps?

2 Consider the following table of costs for the terminal assignment problem. Here the ijth entry indicates the cost of a connection from terminal i to concentrator j.

$$\begin{bmatrix} 4 & 2 & 6 & 8 \\ 3 & 5 & 4 & 9 \\ 3 & 1 & 5 & 10 \\ 9 & 2 & 1 & 9 \\ 11 & 3 & 5 & 12 \\ 5 & 6 & 7 & 18 \end{bmatrix}$$

No more than two terminals may be assigned to each concentrator. Use the optimal algorithm described in Section 3.2 to determine an optimal terminal assignment.

3 A business has several mainframe computers. Terminal concentrators are used to concentrate the traffic from up to 20 terminals over a single channel to one of the mainframes. Write a minimal cost mathematical programming formulation showing all constraints. You must take into account links between terminals and concentrators, links between concentrators and mainframes and the placement of R concentrators.

There are N concentrator sites, M mainframe computers, T terminals, and R placed concentrators. The variables are as follows:

Z_i 1 if concentrator placed at site i (zero otherwise)

S_i cost of concentrator at site i

C_{ij}^2 cost of connecting concentrator i to mainframe computer j

C_{ij}^1 cost of connecting concentrator i to terminal j

X_{ij}^2 1 if there is a connection from concentrator i to mainframe computer j (zero otherwise)

X_{ij}^1 if there is a connection from concentrator i to terminal j (zero otherwise)

4 a. Telephones are to be connected to all digital concentrators. However only some of the telephones are digital. The rest are analog. To connect analog telephone to a digital exchange, an interface box must be purchased for each analog phone. How would a mathematical

program for this concentrator location problem take this requirement
into account?

b. How could one assure, in a mathematical program for a concentrator
location problem, that terminal 17 is *not* connected to concentrators 1,
4, and 7?

5 Consider a concentrator location problem in which there are 20 concentrators in four areas. The first five are in area 1, the next five are in area 2,
the next five are in area 3, and the last five are in area 4.

   Write a mathematical program to ensure that there will be at least one
concentrator selected in each region. There are n terminals. There are m
sites. Each concentrator has a capacity of 15 terminals. Each terminal is
connected to two concentrators. Let the variables be:

C_{ij} cost of connecting terminal i to concentrator j

X_{ij} 1 if terminal i is connected to concentrator j and 0 otherwise

Y_j 1 if concentrator placed at site j, and 0 otherwise

d_j cost of concentrator at site j

6 Consider a concentrator location problem where the data rate of a
terminal is given, in kilobits per second, by

$$X_{ij} = \begin{cases} \text{Cap} & \text{capacity if terminal } i \text{ is connected to concentrator } j \\ 0 & \text{otherwise} \end{cases}$$

Cap can be any positive integer. Each concentrator can handle a
maximum data flow of Q Kbps from the terminals. Each terminal is
connected to one concentrator.

   Write a mathematical program to express this concentrator location
problem. Assume that wiring costs are linear ($C_{ij}X_{ij}$ is the cost of
connecting X_{ij} Kbps from terminal i to concentrator j). If need be,
make use of the function:

$$u(X) = \begin{cases} 1 & \text{if } X > 0 \\ 0 & \text{if } X = 0 \end{cases}$$

7 You must decide where, from a preselected list of locations
$j = 1, 2, \ldots, m$, to place radio base stations that provide radio connections to local stations $i = 1, 2, \ldots, n$.

   The cost of a base station at site j is d_j. If there is indeed a base
station at site j, then $y_j = 1$ (otherwise it is zero). If local station i is in
radio range of the base station at site j, then $x_{ij} = 1$ (otherwise it is zero).

Each terminal must be in range of at least one base station. Each base station can handle up to at least 10 local stations.

Write a mathematical program for this problem. Describe the intention of each equation. Which variables must be solved for? Is there always a solution?

8 Consider a concentrator location problem with n terminals and m sites. Define:

$$X_{ij} = \begin{cases} 1 & \text{if terminal } i \text{ is connected to concentrator } j \\ 0 & \text{if not} \end{cases}$$

Let C_{ij} be the "cost" of connecting terminal i to concentrator j. Let the "cost" be distance in meters. Let D_j be the cost of placing a concentrator at site j.

Let each terminal be connected to exactly two concentrators. The *total* wire used for these two connections must be less than 1000 meters. Each concentrator can service at most 30 terminals.

Write a mathematical program to express this problem formulation.

9 Consider a concentrator location problem. There are m "sites" and n "terminals."

Let C_{ij} be the cost of a connection from terminal i to concentrator j. Let DIS_{ij} be the distance from terminal i to concentrator j. Also let X_{ij} be 1 if terminal i is connected to concentrator j and be 0 if it is not connected.

Let D_j be the cost of purchasing and installing a concentrator at site j. Also let Y_j be 1 if a concentrator is placed at site j and be 0 if it is not.

Write a mathematical program to describe this problem, along with the following additional constraints:

a. The total cost of *connecting* terminals to concentrators is not greater than $C_{wire\$}$.

b. The total cost of purchasing and installing concentrators is no greater than $C_{conc.\$}$.

c. The total length (distance) of wiring used is no greater than DIS_{miles}.

10 Write a mathematical program for the following version of the concentrator location problem. You must account for the cost of:

a. Connecting user i to concentrator j.

b. One interface card in each concentrator for each incoming user line.

c. Placing the concentrators.

Make use of the following variables: Z is the total cost, m is the number of concentrator locations, and n is the number of users.

$$X_{ij} = \begin{cases} 1 & \text{if user } i \text{ is connected to concentrator } j \\ 0 & \text{if not} \end{cases}$$

$$Y_j = \begin{cases} 1 & \text{if a concentrator is placed at site } j \\ 0 & \text{if not} \end{cases}$$

Define any other parameters or variables you'll need.

Also in this problem, there are 100 users. The first 70 are connected to one concentrator each. The rest are connected to two concentrators each. There are 10 potential concentrator locations. The first five can support 20 users. The rest can support 30 users.

11 Write a mathematical program for the following version of the concentrator location problem. In this problem we seek the minimal cost assignment of users to concentrators as well as the placement of concentrators. In this problem it is possible to install one of two types of concentrator at each site (or no concentrator).

	Cost	Capacity
Concentrator A	$1000	10 users
Concentrator B	$3000	50 users

Let Z be the total cost. Let m be the number of concentrator locations, and let n be the number of users. Let C_{ij} be the cost of connecting user i to concentrator j. Finally, let X_{ij} be 1 if user i is connected to concentrator j and be 0 otherwise.

Define any other variables, parameters, or functions you'll need. *Note:* Each user is connected to exactly one concentrator.

12 Consider a network consisting of terminals, concentrators, and super-concentrators. That is, a number of terminals are connected to each concentrator and a number of concentrators are connected to each superconcentrator.

Formulate the design of such a network as two mathematical programs. The first is to determine the placement of concentrators and connect the terminals to concentrators. The second is to determine the placement of superconcentrators and connect the concentrators to the superconcentrators. Each terminal is connected to one concentrator and each concentrator is connected to one superconcentrator. Show the objective functions, constraints (algebraically and in words), and all indexing.

The relevant variables are:

S_i^1 cost of superconcentrator at site i

S_i^2 cost of concentrator at site i

Z_i^1 here $= 1$ if superconcentrator placed at site i

Z_i^2 here $= 1$ if concentrator placed at site i

C_{ij}^1 cost of connecting superconcentrator i to concentrator j

C_{ij}^2 cost of connecting concentrator i to terminal j

X_{ij}^1 here $= 1$ if connection made between superconcentrator i and concentrator j

X_{ij}^2 here $= 1$ if connection made between concentrator i and terminal j

n number of terminals

m number of concentrator sites

m^* number of selected concentrator sites

p number of superconcentrator sites

k_1 maximum number of concentrators that can be connected to superconcentrator

k_2 maximum number of terminals that can be connected to concentrator

13 Randomly place 15 points (nodes) on a piece of paper. Make several copies of this sheet. Run the various spanning tree algorithms discussed (Kruskal, Prim, modified Kruskal, Esau–Williams, and Sharma–El Bardai) on the points on these sheets and compare the results.

14 Run the Ford–Fulkerson–Floyd algorithm on the network of Figure 3.9 with the columns of the table numbered in reverse order.

15 Create a seven-node network with nodes at the following Cartesian grid locations:

Node	Location
1	(0,1)
2	(1,1)
3	(0,0)
4	(1,0)
5	(2,1)
6	(3,1)
7	(4,1)

The following bidirectional links with associated costs exist:

Link	Cost
l(1,3)	1
l(1,2)	2
l(3,4)	1
l(2,4)	1
l(2,5)	10
l(4,5)	9
l(4,6)	5
l(5,6)	1
l(6,7)	1
l(1,7)	2

Run both the Dijkstra and Ford–Fulkerson–Floyd algorithms on this network using node 1 as the root. For the algorithm, number the columns in the table in ascending, and then describing, order.

16 Computer Project Write a computer program to solve the terminal assignment problem using the optimal algorithm discussed in the chapter. You may use any computer language in doing this. The user should be able to input the matrix of terminal assignment costs and the maximum number of terminals to assign to each concentrator. The output should be an optimal assignment of terminals to concentrators. Hand in a report consisting of a program flowchart, the code listing, and examples of the programs use.

17 Computer Project Write a computer program to solve the concentrator location problem using a heuristic approach. Make use of procedures suggested in this chapter such as ADD and DROP. You may use any computer language in doing this. The user should be able to input the matrix of terminal assignment costs, a list of possible concentrator locations, and the maximum number of terminals to assign to each concentrator. The output should be an optimal or near-optimal selection of concentrator sites and assignment of terminals to concentrators. You may use the optimal terminal assignment algorithm of this chapter to make the terminal assignments. Hand in a report consisting of a program flowchart, the code listing, and examples of the programs use.

18 Computer Project Write a computer program to solve the unconstrained minimum weight spanning tree problem using the algorithms discussed in the chapter (Kruskal and Prim). You may use any computer language in doing this. The user should be able to input the node positions. Include an option to have the program generate node positions randomly. The output should be a (perhaps graphical) display of an optimal spanning tree. Hand in a report consisting of a program flowchart, the code listing, and examples of the programs use.

19 Computer Project Write a heuristic computer program to solve the constrained minimum weight spanning tree problem using the algorithms discussed in the chapter (modified Kruskal, Esau–Williams, and Sharma–El Bardai). You may use any computer language in doing this. The user should be able to input the node positions and maximum number of nodes per branch. Include an option to have the program generate node positions randomly. The output should be a (perhaps graphical) display of an optimal or near-optimal spanning tree. Hand in a report consisting of a program flowchart, the code listing, and examples of the program's use.

20 Computer Project Write a heuristic computer program to solve the shortest path routing problem using the Ford–Fulkerson–Floyd algorithm discussed in the chapter. You may use any computer language in doing this. The user should be able to input the node positions and link costs. The output should be a (perhaps graphical) display of an optimal spanning tree. Hand in a report consisting of a program flowchart, the code listing, and examples of the program's use.

4

RELIABILITY THEORY FOR PLANNING

4.1 INTRODUCTION

Our technological civilization depends on the smooth functioning of many machines or systems. A system is composed of many components. The study of reliability is the study of how systems composed of hundreds or thousands of components can function correctly for extended periods of time.

The importance of problems of this type was made apparent in the design of the first electronic computers during and after World War II. These early machines used thousands of vacuum tubes as switching elements. Like light-bulbs, vacuum tubes burn out after a period of use. Keeping the computers with thousands of tubes "up" was a challenge that required good reliability theory. Today such theory is routinely applied to systems ranging from cars to nuclear power plants. It can also be applied to the computer and telecommunication components of systems that are the subject of network planning.

This chapter examines the question of how to compute the "reliability" of networks of components. Here such networks are first viewed statically. That is, it is assumed that the ith component has a constant probability, p_i, of operating correctly (being "up") and a constant probability, $1 - p_i$, of having failed (being "down"). We then want to know that an interconnected network of such components will function—that is, what is the network's **reliability**?

It is possible to examine systems for their reliability dynamically—that is, over time. Lifetime distributions useful for this purpose are outlined in this chapter. Risk assessment, the consideration of the potential negative effects of network failure, is also discussed.

4.2   RELIABILITY OF A NETWORK OF COMPONENTS

4.2.1   Reliability

We define the reliability of a network of components as the probability that the network is "functioning." The reliability of the ith component is the independent probability that it is functioning or p_i. Thus the independent probability that the ith component has failed is $1 - p_i$.

It is assumed here that components are either completely functional or completely inoperative. The reliability of a system of N components is some function of the reliabilities of the individual components:

$$R = f(p_1, p_2, p_3, \ldots, p_N) \tag{4.1}$$

In this section we will first look at three common configurations of components: the series configuration, the parallel configuration, and the k out of N configuration. Then we turn to a number of ways in which more complex configurations can be evaluated.

4.2.2   Series Configuration

In a series configuration all the components must be functional for the overall system to be considered functional. Put another way, the system has failed if at least one component has failed. Then

$$R_{\text{series}} = p_1 p_2 p_3 \cdots p_N \tag{4.2}$$

There is an implicit assumption here that the reliability probabilities of the components are independent of one another. That is, the functioning or failure of any one component neither affect nor depends on that of any other component(s). This independence is what allows us to simply multiply the individual component reliability probabilities together to form R_{series}.

EXAMPLE

If 10 components, each with reliability 0.98, are connected in series, then $R_{\text{series}} = (0.98)^{10} = 0.82$. Note that the effect of configuring N identical components in series is to *lower* the overall reliability. Even though any single component may be very reliable, the ensemble is less reliable.

EXAMPLE

If five components with $p_{1-4} = 0.98$ and $p_5 = 0.6$ are configured in a series manner, then $R_{\text{series}} = (0.98)^4 \times 0.6 = 0.55$. One can see that in a series configuration the overall reliability is only as good as that of the *least* reliable component.

4.2.3 Parallel Configuration

In a parallel configuration the system operates if at least one of the components is functional. The reliability of the overall system is equal to the probability that at least one component is functioning. This is equal in turn to one minus the probability that *all* of the components have failed. Thus

$$R_{\text{parallel}} = 1 - (1 - p_1)(1 - p_2)(1 - p_3) \cdots (1 - p_N) \qquad (4.3)$$

Again, it is assumed that the failure (reliability) probabilities are independent of one another.

EXAMPLE

If four components, each with reliability 0.90, are configured as a parallel system, $R_{\text{parallel}} = 0.9999$. Thus the overall system reliability is *higher* than that of any individual component.

EXAMPLE

If there are five components with $p_{1-4} = 0.6$ and $p_5 = 0.99$ configured in a parallel system, then $R_{\text{parallel}} = 0.9997$. Thus for a parallel configuration, the system reliability is at least as great as that of the *most* reliable component.

4.2.4 *k* out of *N* Configuration

A system may function if at least k out of N components are functioning. Let's assume that the reliability of each component is $p_i = p$. First, an expression for the probability that *exactly* k out of N components are functioning is needed. To get this, we determine that the probability that some *particular* set of k components is functioning is

$$p \cdot p \cdot (\cdots)p(1 - p)(1 - p) \cdots (1 - p) = p^k(1 - p)^{N-k} \qquad (4.4)$$

If this is generalized to the probability that *any* of k out of N components are functioning, one has

$$\binom{N}{k} p^k (1 - p)^{N-k} \qquad (4.5)$$

where $\binom{N}{k}$ represents the number of ways in which k operational components can be selected from the pool of N components. The overall expression can be seen to be the binomial distribution. Now to calculate the reliability of this system, one must determine the probability that k or more components are functioning. This is

$$R_{k/N} = \sum_{i=k}^{N} \binom{N}{i} p^i (1-p)^{N-i} \tag{4.6}$$

EXAMPLE

If six components have $p_i = p = 0.9$ and $k = 4$, then

$$R_{k/N} = \sum_{i=4}^{6} \binom{6}{i} (0.9)^i (1-0.9)^{6-i} \tag{4.7}$$

$$R_{k/N} = \binom{6}{4}(0.9)^4(1-0.9)^2 + \binom{6}{5}(0.9)^5(1-0.9)^1$$

$$+ \binom{6}{6}(0.9)^6(1-0.9)^0 \tag{4.8}$$

$$R = 0.098415 + 0.354294 + 0.531441 = 0.984 \tag{4.9}$$

One can see that the system reliability is higher than that of the individual components. Note that if $k = 1$, one has a parallel system, and if $k = N$, one has a series system. If $1 < k < N$, one has a system reliability that is better than that of a series system but poorer than that of a pure parallel system.

4.2.5 Series and Parallel Configurations

There are many possible configurations of components beyond those mentioned so far. One important class comprises networks of components, where subnetworks consist of components in series or parallel configurations. Anyone familiar with basic electric circuit theory will know that such subnetworks, at least in the electric circuit case, can be replaced by equivalent components. An "equivalent" component is one that can act identically to the subnetwork it replaces with respect to the entire network.

In fact one can use the concept of an equivalent component in reliability theory in much the same way it is used in electric circuit applications. Subnetworks can be replaced by equivalent components, one after another, until an entire network has been replaced by a single equivalent component whose characteristics are the same as that of the network.

To illustrate this, consider the **series–parallel configuration** of Figure 4.1. This network is considered to be functioning if there is at least one path of functioning components from left input to right output.

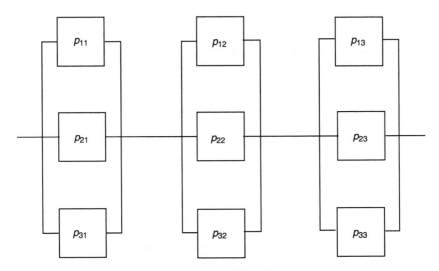

Figure 4.1 Series–parallel configuration.

Each of the parallel subnetworks can be replaced by an equivalent component as follows:

$$R_1 = 1 - (1 - p_{11})(1 - p_{21})(1 - p_{31}) \tag{4.10}$$

$$R_2 = 1 - (1 - p_{12})(1 - p_{22})(1 - p_{32}) \tag{4.11}$$

$$R_3 = 1 - (1 - p_{13})(1 - p_{23})(1 - p_{33}) \tag{4.12}$$

There are now three equivalent components in series, which can be replaced
by a single equivalent component as

$$R_{\text{network}} = R_1 R_2 R_3 \tag{4.13}$$

In general, for a network like that of Figure 4.1 with M elements in each
parallel subnetwork and N parallel subnetworks, we write

$$R_i = 1 - (1 - p_{1i})(1 - p_{2i}) \cdots (1 - p_{Mi}) \qquad i = 1, 2, \ldots, N \tag{4.14}$$

$$R_{\text{network}} = \prod_{i=1}^{N} R_i \tag{4.15}$$

If all $p_{ji} = p$ for $i = 1, 2, \ldots, N$ and $j = 1, 2, \ldots, M$, then [Kapu]

$$R_i = 1 - (1 - p)^M \qquad i = 1, 2, \ldots, N \tag{4.16}$$

$$R_{\text{networks}} = [1 - (1 - p)^M]^N \tag{4.17}$$

As a second illustration, consider the **parallel–series configuration** of Figure 4.2. This network is considered to be functioning if there is at least one path of functioning components from left input to right output.

Each series branch can be replaced by an equivalent component as follows:

$$R_1 = p_{11}p_{12}p_{13} \tag{4.18}$$

$$R_2 = p_{21}p_{22}p_{23} \tag{4.19}$$

$$R_3 = p_{31}p_{32}p_{33} \tag{4.20}$$

One has three components in parallel. Thus,

$$R_{\text{network}} = 1 - (1 - R_1)(1 - R_2)(1 - R_3) \tag{4.21}$$

In general, for a nework like that of Figure 4.2 with N elements in each of M branches, we write

$$R_i = p_{i1}p_{i2} \cdots p_{iN} \qquad i = 1, 2, \ldots, M \tag{4.22}$$

$$R_{\text{network}} = 1 - (1 - R_1)(1 - R_2) \cdots (1 - R_M) \tag{4.23}$$

If all $p_{ij} = p$ for $i = 1, 2, \ldots, M$ and $j = 1, 2, \ldots, N$ then

$$R_i = p^N \tag{4.24}$$

$$R_{\text{network}} = 1 - (1 - p^N)^M \tag{4.25}$$

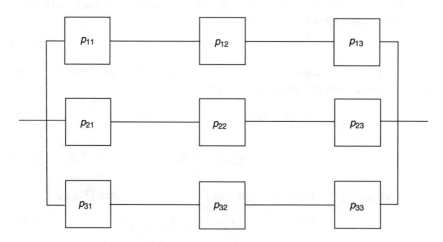

Figure 4.2 Parallel–series configuration.

4.2.6 More Complex Configurations

There are more elaborate configurations of components than the ones discussed so far. Consider Figure 4.3 for instance. This network is considered to be operating (up) if there is at least one connected path from left input to the right output. Clearly, there are no parallel or series subnetworks that can be directly replaced by equivalent components. For a network like this, there are a number of possible solution techniques. Some of these are discussed below. For a look at these techniques from different perspectives, see [Hill 90] [Kapu] [Rao].

Enumeration.   In the solution techniques we shall discuss, each possible state of the system is listed along, with a notation as to whether the system is operational or has failed. For Figure 4.3 this is done in Table 4.1. Here a Boolean logic–like notation is used, where A indicates that component A is in operation and $\bar{A}$ indicates that component A has failed.

The reliability of the system described in Table 4.1 can be found by summing the probability of each mutually exclusive event associated with an operational network. One makes use of all the events that have the network in the "up" state:

$$
\begin{aligned}
R_{\text{network}} = {} & R_A R_B R_C R_D R_E + (1 - R_A) R_B R_C R_D R_E \\
& + R_A (1 - R_B) R_C R_D R_E + R_A R_B (1 - R_C) R_D R_E \\
& + R_A R_B R_C (1 - R_D) R_E + R_A R_B R_C R_D (1 - R_E) \\
& + (1 - R_A) R_B (1 - R_C) R_D R_E + (1 - R_A) R_B R_C (1 - R_D) R_E \\
& + (1 - R_A) R_B R_C R_D (1 - R_E) + R_A (1 - R_B)(1 - R_C) R_D R_E \\
& + R_A (1 - R_B) R_C (1 - R_D) R_E + R_A (1 - R_B) R_C R_D (1 - R_E) \\
& + R_A R_B (1 - R_C)(1 - R_D) R_E + R_A R_B (1 - R_C) R_D (1 - R_E) \\
& + R_A (1 - R_B)(1 - R_C) R_D (1 - R_E) \\
& + (1 - R_A) R_B (1 - R_C)(1 - R_D) R_E
\end{aligned}
$$

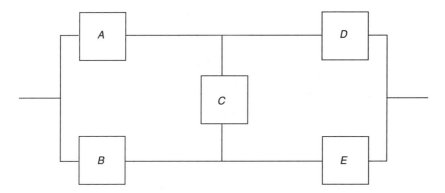

Figure 4.3 A complex configuration.

Table 4.1 Enumeration of the States of a Sample System

Number of Component Failures	Event	System Condition
0	1. $ABCDE$	Up
1	2. $\overline{A}BCDE$	Up
	3. $A\overline{B}CDE$	Up
	4. $AB\overline{C}DE$	Up
	5. $ABC\overline{D}E$	Up
	6. $ABCD\overline{E}$	Up
2	7. $\overline{A}\overline{B}CDE$	Down
	8. $\overline{A}B\overline{C}DE$	Up
	9. $\overline{A}BC\overline{D}E$	Up
	10. $\overline{A}BCD\overline{E}$	Up
	11. $A\overline{B}\overline{C}DE$	Up
	12. $A\overline{B}C\overline{D}E$	Up
	13. $A\overline{B}CD\overline{E}$	Up
	14. $AB\overline{C}\overline{D}E$	Up
	15. $AB\overline{C}D\overline{E}$	Up
	16. $ABC\overline{D}\overline{E}$	Down
3	17. $\overline{A}\overline{B}\overline{C}DE$	Down
	18. $\overline{A}\overline{B}C\overline{D}E$	Down
	19. $\overline{A}\overline{B}CD\overline{E}$	Up
	20. $\overline{A}B\overline{C}\overline{D}E$	Down
	21. $\overline{A}B\overline{C}D\overline{E}$	Down
	22. $\overline{A}BC\overline{D}\overline{E}$	Down
	23. $A\overline{B}\overline{C}\overline{D}E$	Up
	24. $A\overline{B}\overline{C}D\overline{E}$	Down
	25. $A\overline{B}C\overline{D}\overline{E}$	Down
	26. $AB\overline{C}\overline{D}\overline{E}$	Down
4	27. $\overline{A}\overline{B}\overline{C}\overline{D}E$	Down
	28. $\overline{A}\overline{B}\overline{C}D\overline{E}$	Down
	29. $\overline{A}\overline{B}C\overline{D}\overline{E}$	Down
	30. $\overline{A}B\overline{D}D\overline{E}$	Down
	31. $A\overline{B}\overline{C}\overline{D}\overline{E}$	Down
5	32. $\overline{A}\overline{B}\overline{C}\overline{D}\overline{E}$	Down

Naturally one can make the association $R_A = p_1, R_B = p_2, R_C = p_3, R_D = p_4$ and $R_E = p_5$. This associates probabilities with each variable.

Using techniques from Boolean algebra, expressions of this type can usually be simplified. The equation above is equivalent to

$$R_{\text{network}} = R_B R_E + R_A R_D + R_A R_C R_E + R_B R_C R_D \qquad (4.26)$$

This method is robust and allows a detailed exposition of the system states. The difficulty, naturally, is that the number of states grows exponentially with the number of components. For N components, there are 2^N possible states. Thus this technique is computable only for relatively small numbers of components (say around 25 or less).

More specifically, if there are N components with exactly M possible failures, the number of failure modes is $\binom{N}{M}$.

Thus the total number of possible events is given by

$$\text{number of events} = \sum_{M=0}^{N} \binom{N}{M} = 2^N \tag{4.27}$$

4.2.7 EXACT CALCULATION OF LARGE SYSTEMS

Techniques applied to larger models than can be solved by enumeration makes use of graph theoretic concepts of minimal cuts and minimal paths. A **minimal path** is a minimal set of components in a network of components that, if all the components in the set are operational, guarantee that the entire network is operational. Conversely, a **minimum cut** is a minimal set of components in a network of components that, if all the components in the set have failed, guarantee that the network has failed.

As noted in [Hill 90], when minimal paths are used there are two methods by which the exact reliability of a network may be calculated. In the first method, the probability of the union of the events where each of the minimal paths is operational is computed. Again, we use a Boolean logic–like notation, where A indicates that component A is operational.

For the example of Figure 4.3 one has

$$R_{\text{network}} = \text{Prob}[(BE = 1) \cup (AD = 1) \cup (ACE = 1) \cup (BCD = 1)] \tag{4.28}$$

One can use Boolean algebra to find an algebraic expression from this description that is equivalent to equation (4.26).

In the second method, one treats each path as a single component in a parallel system. Thus for the example of Figure 4.3, one deals with the logical expression known as the "structure function" (S.F.) [Kuma] [Hill 90]:

$$1 - (1 - BE)(1 - AD)(1 - ACE)(1 - BCD) \tag{4.29}$$

Simplifying by means of Boolean algebra, one can obtain an expression equivalent to (4.26). To do this one must realize that

$$R_{\text{network}} = \text{Prob}[\text{S.F.}(A, B, C, D, E) = 1] \tag{4.30}$$

One can also use minimal cuts in more than one way [Hill 90] to calculate the exact system reliability. One approach is to compute one minus the probability of the union of all the events where minimal cuts cause system failure. So one has

$$R_{\text{network}} = 1 - \text{Prob}[(A = 0, B = 0) \cup (D = 0, E = 0)$$
$$\cup (A = 0, C = 0, E = 0) \cup (B = 0, C = 0, D = 0)] \quad (4.31)$$

One can then use Boolean algebra to compute an algebraic expression equivalent to equation (4.26).

A second approach to computing system reliability by means of minimal cutsets is to write the logical expression (structure function):

$$[1 - (1 - A)(1 - B)][1 - (1 - D)(1 - E)][1 - (1 - A)(1 - C)(1 - E)]$$
$$\times [1 - (1 - B)(1 - C)(1 - D)] \quad (4.32)$$

Here each bracketed term is 1 if at least one component in the associated cut is operational. Thus the overall logical expression has a value of 1 (system operational) if at least one element in each minimal cut is operational. Boolean algebra can be used again to transform the expression above into one similar to equation (4.26).

Boolean algebraic solution techniques, which are very well developed for such fields as digital circuit design, also can be used for the reliability problem. Karnaugh maps (see any book on digital circuit design: e.g., [Taub]) in particular provide a simple-to-use graphical solution technique for smaller size problems. For instance, by enumerating all the basic events on a Karnaugh map (or set of them), one can group 1s or 0s to simply determine the simplest possible algebraic expression for a particular model.

4.3 RELIABILITY AND LIFETIMES

4.3.1 Introduction

Reliability, in the context of Section 4.2, could be understood with a minimal background in probability and statistics. In fact though, the complete theory of reliability relies heavily on these subjects. A most important concept in this field is that of **lifetimes** [Leem]. A component lifetime is the period during which a component functions properly. A system lifetime is the time period during which a system of components functions properly. Reliability engineers are interested in calculating such quantities as the lifetime of a system of components.

Actually the mathematics used in reliability engineering is much more general than this, and can be applied to other fields. For instance, actuaries, who generally work for insurance organizations, are interested in the "lifetimes" of individuals, particularly as this variable applies to the profitable underwriting of insurance policies. This mathematics is also applicable to the profitable design of warranty policies [Elsa]. On the other hand, statisticians who work in the biological sciences are interested in studying and making predictions on the lifetimes of organisms. This could include patients receiving certain (perhaps experimental) treatments.

This section contains a tutorial treatment into reliability statistics useful in reliability engineering in general and in network planning in particular. The interested reader who seeks to work professionally with these techniques can refer to a number of excellent texts on the subject [Elsa] [Kuma] [Leem].

4.3.2 Lifetime Measures

Section 4.2 discussed reliability at a single point in time. There are a variety of ways to portray reliability as a function of time. Generally these take the form of what are called lifetime distributions. Three of these, which apply both to continuous (e.g., a fuse) and discrete (e.g., an emergency generator that is tested weekly) lifetimes, are discussed below. The interested reader is referred to Leemis [Leem] for a more complete discussion of such lifetime distributions.

Survivor Function.   Following the notation and nomenclature in [Leem], let's introduce $S(t)$, the "survivor" function $S(t)$, which is a function of time t, can be defined for a component as the probability that the component is operational at time t. It can also be defined as

$$S(t) = 1 - F(t) \tag{4.33}$$

where $F(t)$ is the cumulative distribution function of failure.

If there are many independent and identically distributed (i.i.d.) components, the survivor function is simply the expected fraction of operational components at time t.

With this definition in mind, it can be seen that usually $S(t = 0) = 1.0$ and in the absence of repairs $S(t)$ generally is monotonically nonincreasing. Thus $\lim_{t \to \infty} S(t) = 0$. The survivor function is sometimes called the "reliability" function.

Probability Density Function.   The lifetime **probability density function** is $f(t)$. Naturally it can be seen that it is also a function of time. It is a probability density function like those in other fields. Thus $f(t) \geq 0$ for all $t \geq 0$ and $\int_0^\infty f(t)dt = 1$. The *probability* that the failure time for a component is at time T, where $t_1 \leq T \leq t_2$ is

$$\int_{t_1}^{t_2} f(\tau)d\tau \tag{4.34}$$

The probability density function will be larger in value when failure is more likely and smaller in value when failure is less likely. The probability density function can be related to the survivor function as follows:

$$S(t) = 1 - F(t) \tag{4.35}$$

$$S'(t) = -f(t) \tag{4.36}$$

$$f(t) = -S'(t) \tag{4.37}$$

where the prime indicates the first derivative with respect to time.

Since $S(t)$ is equal to $1 - F(t)$:

$$S(t) = \int_t^\infty f(\tau)d\tau \tag{4.38}$$

Hazard Function.   The **hazard function**, $h(t)$, is a popular way of representing the "risk" of a component failing at time t. Some refer to it as the failure rate or hazard rate [Leem].

The hazard function can be stochastically interpreted for small Δt as

$$h(t)\Delta t = \text{Prob}[t \leq T \leq t + \Delta t | T \geq t] \tag{4.39}$$

where T is the failure time. In words, one is saying that the hazard function, over a small interval, is proportional to the probability of failure given that the failure time is greater than or equal to t: that is, given that the component has lasted until time t.

Let's make the origin of the foregoing definition somewhat clearer.

From what we know of survivor and probability density functions, we can write

$$\text{Prob}[t \leq T \leq t + \Delta t] = \int_t^{t+\Delta t} f(\tau)d\tau \tag{4.40}$$

$$\text{Prob}[t \leq T \leq t + \Delta t] = \int_t^\infty f(\tau)d\tau - \int_{t+\Delta t}^\infty f(\tau)d\tau \tag{4.41}$$

$$\text{Prob}[t \leq T \leq t + \Delta t] = S(t) - S(t + \Delta t) \tag{4.42}$$

Now assuming that the component in question was operational at time t and using Bayes formula, the foregoing relations becomes

$$\text{Prob}[t \leq T \leq t + \Delta t | T \geq t] = \frac{\text{Prob}[t \leq T \leq t + \Delta t]}{\text{Prob}[T \geq t]} \tag{4.43}$$

$$\text{Prob}[t \leq T \leq t + \Delta t | T \geq t] = \frac{S(t) - S(t + \Delta t)}{S(t)} \tag{4.44}$$

The hazard function is the limit of this differential-like function averaged over time; that is,

$$h(t) = \lim_{\Delta t \to 0} \frac{S(t) - S(t + \Delta t)}{S(t)\Delta t} = \frac{-S'(t)}{S(t)} \tag{4.45}$$

These equations tell us how the original definition that started this section was arrived at

$$h(t) = \frac{-S'(t)}{S(t)} = \frac{f(t)}{S(t)} = \frac{f(t)}{\int_t^\infty f(\tau)d\tau} \tag{4.46}$$

The equalities above relate the survivor and probability density function to the hazard function.

The units of the hazard are failures per unit time. Thus one might have values such as 2.3 failures per minute or 1.7 failures per year.

The hazard function for a negative exponential probability density function is somewhat unique among all continuous probability density functions:

$$f(t) = e^{-\lambda t} \tag{4.47}$$

$$h(t) = \frac{e^{-\lambda t}}{\int_t^\infty e^{-\lambda \tau}d\tau} \tag{4.48}$$

$$h(t) = \frac{e^{-\lambda t}}{-(1/\lambda)[0 - e^{-\lambda t}]} \tag{4.49}$$

$$h(t) = \frac{\lambda e^{-\lambda t}}{e^{-\lambda t}} = \lambda \tag{4.50}$$

That is, the negative exponential probability density function is associated with a *constant* hazard function. It intuitively means that the risk of the component failing does not change with the operational age of the component. This is actually a consequence of the memoryless nature of the negative exponential density [Klei 75] [Robe 94]. Intuitively, from the memoryless property, one recognizes that a knowledge of the past history of negative exponentially distributed random variable will not increase the accuracy with which the future can be predicted.

One can construct a Markov "failure process" as a series of failure events with a negative exponential distributed amount of time between each. If the rate of failure is constant over time, then any point in time is as likely to be a failure event as any other. This is true even if we know the history of the failures up until time t_0. Knowing this, it is not too surprising that the hazard function of a negative exponential density is constant—the risk is the same no matter at what time instant is at issue and regardless of how complete our knowledge of the past history.

Just as with the probability density function, the larger the value of the hazard function, the larger the probability of failure. However, whereas the probability density function directly expresses probability, the hazard function directly expresses "risk."

There are several possible shapes for the hazard function. A common one, the **bathtub function,** has a large initial hazard value that decreases to a small constant value for an extended period of time and then increases after some point. The bathtub function is particularly relevant to manufacturers. Mechanical/ electronic systems often have an initial burn-in period when failures due to manufacturing defects occur at a relatively high probability. Electronic components are sometimes operated in the factory to eliminate these failures prior to sale. The burn-in period is followed by an extended period that comprises most of the useful life of the component with a low, constant risk of failure. Finally, at some point components wear out and the hazard (risk) increases accordingly.

Other shapes of the hazard function are possible. The hazard function may be increasing over the lifetime, indicating constant wear-out. In some specific cases (e.g., a complex software system in which programming errors are detected with time and fixed), the hazard function may decrease over the system lifetime.

Other lifetime distributions of interest include the cumulative hazard function and mean residual life function. The reader is referred to [Leem] for a more complete treatment of lifetime distributions.

4.4   PROBABILISTIC RISK ASSESSMENT

There is a large body of work [Kuma] dealing with the assessment of the reliability of complex systems through probabilistic models. This is done with the aim of performing **risk assessment**. Risk assessment deals with the negative consequences of system or component failure. In the past these have typically been applied to the nuclear, chemical, and transportation industries, where any system failure has serious potential safety and environmental consequences. However the increasing sophistication of computer and telecommunication networks and their key role as the "nervous system" of complex and crucial systems indicate that risk assessment techniques may be very relevant to the study of network planning.

Two mainstays of risk assessment are event trees and fault trees.

In an **event tree** an "initiating event" is associated with the root [Kuma]. This initiating event may represent an accident or a component failure. Further potential events (e.g., power failure) are associated with each level of the tree. Whether a part of the tree "branches" in each case generally depends on whether an associated event does or does not occur. Associated with each terminal leaf of the tree is a probability that is computed by considering the path from the root to the terminal leaf and the probabilities of each branching along the path. The sequence of events constituting the path is known as an **accident sequence**.

Fault trees are somewhat different. Boolean logic is used to construct a logic gate type schematic, or written logic expressioins, in which potential events are the inputs and the accident whose probability of occurrence needs to be studied is the final (usually single) output. Thus the fault trees tries to capture such relations as (IF THE POWER FAILS) AND (THE BACKUP POWER FAILS) THEN (THE SYSTEM IS DOWN). Alternatively, a relation might be (IF UNINTER-RUPTIBLE POWER SUPPLY 1 FAILS) OR (IF UNINTERRUPTIBLE POWER SUPPLY 2 FAILS) THEN (THE SYSTEM HAS PARTIAL POWER). Fault trees allow a hierarchical consideration of such conditions and outputs.

Event trees as often said [Kuma] to embody forward or inductive logic. They are used in an attempt to seek ramifications of some basic (initiating) event. That is, they are used to answer such questions as "What happens if the power supply fails?" or "What happens if the fiber optic link is cut?" Fault trees are said to embody backward or deductive logic. They are used in an attempt to find the likely cause of significant events. That is, they are used to answer such questions as "How could the power supply fail?" or "How could the fiber optic link fail?"

Failure events can involve human error, computer error, design error, or maintenance error [Kuma]; or they can be failures of hardware, software, or the environment (e.g., earthquakes). Even trees and fault trees allow one to schematically capture basic relationships between faults and events. They allow the analyst to abstract out key failures and guide quantitative/qualitative system reliability analysis [Fuss].

Computer and telecommunication networks today act as the nervous system for many critical systems. In a single day, for example, a stock brokerage may lose commissions worth many millions of dollars if its networks are not functioning. The same sort of potential for economic loss holds for banks, catalog companies, and factories. Thus, in planning the evolution of such networks over time, reliability and risk assessment should be considered. Such a consideration may lead to an evaluation of the robustness of equipment and software, backup systems, standby power, connectivity, and the support ability of network management tools.

4.5 PROBLEMS

1 At least how far back does the use of reliability theory go?

2 If five components, each with reliability 0.9, are placed in series, what is R_{series}?

3 Why is it that in a series configuration of components, the ensemble is less reliable than the individual components? Why is it that the ensemble

reliability is less reliable that the reliability of the least reliable component?

4 If five components of reliability 0.9 are placed in parallel, what is $R_{parallel}$?

5 Why is it that in a parallel configuration of components, the ensemble is more reliable than the individual components? Why is it that the ensemble reliability is more reliable than the reliability of the most reliable component?

6 Compute $R_{k/N}$ for five components if $p_i = p = 0.9$ and $k = 3$.

7 Why is it that the reliability of a k/N configuration is between that of a series and a parallel configuration of the same size and parameters?

8 Consider a series–parallel configuration where they are $N = 4$ subnetworks in series of $M = 4$ parallel components each. Let $p = 0.9$. Find $R_{network}$. Relate $R_{network}$ to that of a series network of four components where $p = 0.9$.

9 Consider a parallel–series configuration where there are $N = 4$ components in each branch and $M = 4$ branches. Let $p = 0.9$. Fiind $R_{network}$. Relate $R_{network}$ to that of a parallel network of four components where $p = 0.9$.

10 Consider a system with three parallel links between two routers. Let the independent reliability of each link be p. What is the smallest value of p that achieves an overall system reliability of 0.999? Do *not* consider the reliability of the routers. How valid is the independence assumption if all three cables use the same cable duct?

11 A private network has two parallel links to a gateway, each with reliability p. The gateway has reliability q. A single link leads from the gateway to the outside world. It has reliability r. The reliabilities of the component are mutually independent. Find an expression for the reliability of the overall access system. Do *not* consider the reliability of the private network itself.

12 Consider three telecommunication switches, S1, S2, and S3. There are four parallel links of 100 Mbps each between S1 and S2 (left hop). There are also four parallel links of 70 Mbps each between S2 and S3 (right hop).

   The reliability probabilities of the three switches are p_1, p_2, and p_3, respectively.

   The reliability probabilities of the left hop links are p_4–p_7, respectively.

   The reliability probabilities of the right hop links are p_8–p_{11}, respectively.

   The reliabilities of the components are mutually independent.

Draw a diagram of the system. Calculate the reliability of the overall system if it is assumed to be functioning if there is at least 200 Mbps of capacity from end to end (from S1 to S3). Let the reliability probabilities of the left hop links be p and that of the right hop links be q.

13 Consider two telecommunication switches, A and B, with four paths between them. The first path consists of a direct link. The second path passes through switch C (there are two links, AC and CB). Similarly, the third path passes through switch D. Finally, the fourth switch passes through switches E and F (thus creating links AE, EF, and FB).

The direct link has reliability q. All other links each have reliability of p. Assume that the switches are always "up."

Draw a diagram of the system. The system is considered to be functioning if there is at least one path from switch A to switch B. Calculate the reliability of the system. Show all work.

14 Consider a video network carrying signals between nodes A and B. There are two paths between A and B, one through node C (upper path), and one path through node D (lower path). There are three parallel 100 Mbps links between each node pair: AC, CB and AD, DB. The independent reliability of each link is p. Draw a diagram of the system. A video signal is to be transmitted from node A to node B. The video signal requires 200 Mbps. Because of technological limitations all of this 200 Mbps must follow the same path. That is, the signal is unreliable if it is split between the upper (ACB) and lower (ADB) paths. Thus the system is considered to be functioning if there is at least 200 Mbps of capacity on *either* the upper or lower path or on both. Determine the reliability of the system. Show all work.

15 Consider a fiber optic network in the form of a binary tree. The root is in Toronto. There are three levels and eight terminal nodes. The three rightmost terminal nodes are in the Boston metropolitan area and the five leftmost nodes are in the New York City metropolitan area. The reliability of each of 14 links $(2 + 4 + 8)$ in the binary tree is p. Assume that all the nodes are always "up." Draw a diagram of the system. The Toronto–Boston link is considered to be operating (reliable) if there is at least one functioning path from Toronto to Boston. Calculate the reliability of the Toronto–Boston connection.

Are the reliabilities of transmissions into the New York and Boston areas independent of each other? Briefly explain in two or three sentences.

16 Compute $R_{network}$, as is done in Section 4.2.6, for the following network [Elsa]. Once again, the network is assumed to be functioning if there is at least one path (from node 1 to node 4). Note that all paths in the network of Table 4.2 are unidirectional. Be sure to provide a simplified expression as in equation (4.26).

TABLE 4.2 NETWORK FOR PROBLEM 16

Component	Between
A	Input node 1 and node 2
B	Node 2 and node 3
C	Node 3 and output node 4
D	Input node 1 and node 3
E	Node 2 and output node 4

17 For N components there are 2^N states. Find the number of states when $N = 5, 10, 20, 30, 50, 100$. What do these values suggest to you?

18 For the network of Problem 16, find $R_{network}$ using the concept of minimal paths. Do the same using the concept of minimal cuts.

19 Consider the Weibull distribution family:

$$S(t) = e^{-(\lambda t)^{\kappa}}$$

Find $f(t)$ and $h(t)$ [Leem].

20 Find the survivor function, $S(t)$, as a function of the hazard function, $h(t)$.

21 Find the probability density function, $f(t)$, as a function of the hazard function, $h(t)$.

22 For what choice of $f(t)$ is $h(t)$ constant? What is the significance of a constant $h(t)$?

23 What is the significance of the bathtub function?

24 What is risk assessment? Why is it relevant to network planning?

25 What is an event tree? How is it used? What type of logic is it said to embody?

26 What is a fault tree? How is it used? What type of logic is it said to embody?

27 Consider the questions "What happens if the satellite link is lost?" and "How could the satellite link be lost?" Which of these questions would be studied by means of events trees and which by means of fault trees? Why?

CHAPTER 5

SOFTWARE AND OPTIMIZATION FOR PLANNING

5.1 INTRODUCTION

The single biggest breakthrough in network planning over the past 20 years has been the availability of increasing power for computers in general and software in particular. Naturally, this growth was part of a broad trend that has aided the development of many fields, including network planning. This increase in hardware and software power has led to the automation of planning data, computation, and recommendations. Thus the first half of this chapter deals with such key computer technology developments for planning such as object-oriented programming, database systems, geographic information systems, and expert systems.

The increasing power of hardware and software power has also led to new approaches in optimization that could not be implemented earlier. Many of these are inspired by novel concepts from other fields. Thus the second half of this chapter discusses classical nonlinear model optimization but also tabu search (inspired by human intuition and memory), simulated annealing (inspired by metallurgy), genetic algorithms (inspired by evolutionary biology), and neural networks (inspired by the brain).

This is an exciting time period to be working in any one of these areas. Hopefully this chapter conveys some of that excitement.

5.2 OBJECT-ORIENTED PROGRAMMING

5.2.1 Introduction

As complex software systems were developed during the 1970s and 1980s it was apparent that there were real problems with creating and using such systems. The development of complex software was extremely labor intensive, hence very costly. Moreover the maintenance of such code over the years of its use was labor intensive and expensive.

An appreciation grew that through well-thought-out, systematic design, these problems could be mitigated. The main vehicle for such systematic design is what is called **object-oriented programming**.

Object-oriented programming is a programming methodology with two key aspects. The first is that the actual features of the environment the software is being designed for are embedded in the software. The second aspect is that the minutia of complex software implementation are hidden.

Actually, as Cusack and Cordingley [Cusa] point out, there is no single universally accepted object-oriented methodology. There is, in fact, no single object-oriented programming language. What does exist is a body of basic ideas that can be tailored to user needs. These basic ideas are outlined next [Coad] [Cusa] [Meye] [Pohl] [Stro].

5.2.2 Objects, Class, and Type

An **object** is a representation of a distinct element in an overall system. The system may be, for example, a corporate network. Typical objects include elements such as particular routers, gateways, PCs, workstations, fiber optic cables, and coaxial cables. It is important to realize that these are distinct elements with obvious boundaries.

We can in a general way describe (i.e., specify or program [Cusa]) an object such as a router without going into the details of a particular vendor's implementation. A general description of a related group of objects is called a **class**. An object is an **instance** of the class of which it is a member.

Thus, object-oriented programming has a data-oriented outlook on the world where data (state) and behavior (process) are closely tied together [Pohl]. That is "class" represents some particular data and some particular behavior and objects are the "instances" of class.

A concept related to class is that of "type." The **type** of an object is a limited description of its attributes. For instance, many, but not all, PCs are Windows-based. Naturally, a type can be identified for each class, and a type specifies a related group of objects. In fact the usual object-oriented methodologies/ languages will have the concept of either class or type, but not both.

Subtyping refers to classes that "contain" other classes. For instance, both the classes `coaxial cable` and `fiber optic cable` are contained in the class `link`. Some classes may be subtypes of more than one other class. One reason for the usefulness of subtyping is that a subtype object can be placed in an environment expecting the supertype object.

5.2.3 Encapsulation

Encapsulation refers to the idea that the internal operation of each object is not apparent. The router routes, the PC computes, the fiber optic link carries data, and so on. An object **interface** is the object's specified manner of interacting with its surroundings. While the interactions may change the state (data) of an object, that object does not change its **identity**. Encapsulation is useful for making a "clean" software design and for making an object implementation inaccessible to client code [Cord] [Pohl].

Note that **active objects** may initiate interactions with their environment, whereas **passive objects** do nothing until an action is requested of them.

5.2.4 Inheritance

Inheritance refers to the process of describing a new type of object in terms of modifications to an existing type [Cusa]. Put another way, new classes are "derived" from established classes [Pohl]. In the case of **single inheritance** there may be one parent type. In the case of **multiple inheritance** there may be several parent types.

An inherited type may have a modified set of interactions compared to its parent type and thus may not act as a substitute for the parent type. It is also not a subtype. In the same way, a parent type is not necessarily a supertype.

5.2.5 Object-Oriented Programming Advantages

Object-oriented programming is very useful for complex software projects that require the coordinated work of several programmers. Beyond this the "real-world" nature of objects and their relationships improves understanding and coordination among system analysts, application experts, and programmers. The programmers can work on individual code sections independently and in a way that preserves the integrity of the overall code.

Initially, writing software in an object-oriented style requires somewhat more discipline and effort than not doing so. However, the payoff comes as the software ages and must be maintained and modified. The original software requires much less change, and such change is likely to be controlled and localized. Code reuse is also much easier. This is a marked contrast to "legacy" software written prior to the advent of object-oriented programming. Whereas such software is often too valuable to stop using, it often is too expensive to rewrite using object-oriented concepts. Being implemented in pre-object-oriented languages such as COBOL, legacy programs are exceedingly costly to maintain and modify.

5.2.6 C++

Among object-oriented languages are C++ [Pohl] [Stro], Smalltalk [Gold 83], and Eiffel [Meye]. Among the most popular today is C++, a superset of the older C language. A C program will compile on a C++ compiler. C was created by D. Ritchie in the early 1970s as a system implementation language. The UNIX operating system was constructed using C.

C++ was developed by B. Stroustrup in the early 1980s. He incorporated into C certain object-oriented constructs, as in the earlier language Simula 67 to create C++, which was intended to be a language for the serious programmer. C++ uses object-oriented concepts but has the elegance and wide acceptance associated with C.

The compatibility of C++ with the popular C aided the acceptance of the new language. In addition, C++ does not need a graphics environment, and the use of strong typing makes it safer than C [Pohl].

5.2.7 Conclusion

The use of object-oriented programming techniques has become very widespread. There are many books available on object oriented programming. A 1993 article by Cusack and Cordingley [Cusa] provides a concise, well-written tutorial.

5.3   DATABASES

5.3.1 Introduction

Much of the computer revolution that is invisible to the public involves our increasing ability to create, manipulate, and search very large amounts of information. Computer sysetms that do this, known as **databases**, can be implemented on any platform, including PCs, workstations, minicomputers, and mainframes. A variety of software specifically designed for database systems is available.

Databases have found widespread use in a large variety of fields, including network planning. Among the uses of databases are querying (asking questions about information), providing reports (receiving answers to queries), creating graphical summaries of data, discovering unexpected patterns in a database's information, performing quality maintenance of the data store, and merging data from different sources (i.e., data fusion).

5.3.2 Database Models

There are five main types of model for the logical structure of databases [Pars] [Sale].

The simplest model, and the least powerful, is that of **file search systems**. In this model data is typically stored in one file. A complete search of the file is performed when data is needed. This serial searching limits the throughput (i.e., number of searches per second) of such systems, particularly when there are very large amounts of data and large numbers of users.

Another possible model is that of a **hierarchical database**. Here data is organized in a tree-type structure. Each level of the three-type data structure is associated with a different type of data. For instance, a tree will have a root (usually associated with the database itself). In a network planning application, there may be a level for device classes (e.g., routers, switches, links, platforms) below the root. Below the level of device classes there may be a level for particular devices. Below that level may be a level for device parameters, and on and on. Finding the throughput of a switch in a hierarchy does not require a search of the entire database, as in a file search system.

Hierarchical databases can perform well when appplications fit the tree model. Problems with the hierarchical model include lack of flexibility, the need to set up the tree structure when setting up the database, and complications associated with changing the levels of data and data items that appear at several places in the tree.

In a **network database**, model data can be thought of as occupying nodes in a logical network with links (pointers) possible between any of the nodes. Thus duplicated data, as discussed above for hierarchical databases, ceases to be a problem. Data can in fact be arranged in sets with links between the different elements of each set. Accordingly, the main advantage of the network model is its ability to capture complex relationships. But this power comes at a price. Relationships must be mapped out by the database implementors. Changing the logical structure, once it has been implemented, may be difficult.

Mainframe databases up until the early 1970s used largely hierarchical or network models. Then E. F. Codd at IBM introduced a new model called the **relational database**. It became the most popular database model.

Data in relational databases is stored in two-dimensional tables. Each (horizontal) row is a distinct item, while each (vertical) column contains **attributes** of the data. Each table (relation) has a name. Table 5.1 is an example table of a hypothetical instance of the relation SWITCH.

Updates to a relational database are made by "cutting and pasting" [Pars] the rows and columns in tables. Languages for relational databases use either **relational algebra** or **relational calculus**. In relational algebra, specific operators (e.g., selection, projection, product, union, set difference, join) are used to manipulate relationships. [Hugh] [Pars]. Relational calculus allows the expression of queries.

Relational databases provide a high degree of flexibility in defining relationships, need no pointers to data as in the hierarchical and network models, and are well suited for transaction processing. Moreover, the underlying model of two-

Table 5.1 Example Table for a Relational Database

ID #	Manufacturer	# Ports	Location	ISDN?
1	Acme Inc.	100	Bldg. 33	No
2	Acme Inc.	1000	Bldg. 47	No
3	Beta Inc.	1024	Bldg. 12	Yes
5	Delta Switch Co.	128	Bldg. 9	Yes
7	Beta Inc.	2048	HQ	Yes
9	Delta Switch Inc.	128	Bldg. 9	Yes

dimensional tables is very intuitive. This property helped the relational model to achieve widespread acceptance. There are indeed some problems with the relational model, however, including the handling of user-defined data types [Sale].

The final, and most recent approach to database modeling is that of **object-oriented databases**, which apply the object-oriented approach of programming to database information. Despite the surface appeal, this approach has its own set of problems [Sale]. Parsaye and Chignell [Pars] suggest a relational model at the lower level with an overlay of an object model for user functionality.

5.3.3 Database Architectures and Platforms

A number of different system architectures and platforms for databases are possible. In **centralized systems** a large computer such as a mainframe or minicomputer serves as a hub, and many relatively "dumb" terminals are wired into it. A mainframe can support over a thousand users concurrently, while a minicomputer may support in the low hundreds of concurrent users. Mainframes are more secure and reliable than PC systems (see below). However, they are expensive to maintain, requiring dedicated staffs and special facilities.

In a **PC database** the database management software and database application software may reside on a single PC. Alternatively, if a number of PCs are networked through a local area network, the management software may reside on a single PC (file server). The application software residing on other PCs can retrieve files from the file server for processing. In such a **multiuser environment**, a system of "locks" is usually used to ensure that when one PC has a file it is updating, other PCs cannot make changes to the file during the updating.

Most PC-based databases use the relational model (or in some cases file search systems). The total number of users supported by PC database systems is limited compared to centralized systems because of the slower speed of PCs and potential network congestion.

A popular computing paradigm that has been applied to databases is that of the client/server architecture. In a **client/server database** the database system is separated into a database application that resides on "client" PCs and the database management system, which resides on a database server. The **front-end** database application software is responsible for screen presentation and I/O. The **back-end** database server is responsible for the data manipulation, retrieval, and storage. PCs are the usual platform of choice for client/server architectures.

Among the advantages of client/server architectures are reduced networked traffic (compared to the basic PC systems that ship files back and forth), enabling the use of a wide variety of PCs as front ends and enabling a high degree of data integrity at the back end, including possible encryption of data. Disadvantages of client/server architectures include the cost of support staff, hardware, and software compared to some of the more basic database solutions [Sale].

Finally, in a **distributed processing database** system, information is stored on a variety of interconnected platforms. A request for information will cause the platform accessed to either deliver that information directly or obtain it over a network from another host. Distributed processing is a new area, and such questions as preserving data consistency and integrity continue to receive attention.

5.3.4 Database Software

The importance of the software component of a database system cannot be overestimated. Among software engineering issues that come up in the design of database systems are data abstraction, concurrency, and the correctness of programs [Hugh]. Database languages include common procedural languages such as COBOL and C, the popular Structured Query Language (SQL) for relational models, and object-oriented languages such as C++. A new trend, "intelligent databases," seeks to maximize the utility of information through appropriate information visualization, discovery, interpretation, and presentation [Pars].

5.3.5 Conclusion

Network planning involves a knowledge of extensive communication facilities and large numbers of users. Thus it is not too surprising that modern databases holding both network and user information play a vital role in this enterprise.

The interested reader can find more detailed information on databases in the *IEEE Transactions on Knowledge and Data Engineering* and in the *ACM Transactions on Database Systems*.

5.4 GEOGRAPHIC INFORMATION SYSTEMS

5.4.1 Introduction

In this section we discuss one of the fastest growing applications of computer power in recent years: **geographic information systems (GIS)**. These are computer systems that combine the ability to store, display, and reproduce spatial data (i.e., maps) with an ability to analyze spatial data. This ability to provide an analysis of data distinguishes GIS from computer-assisted mapping (CAM), which uses CADD (comuter-aided design and drafting) technology to simply computerize maps. Also, a GIS's storage of spatial relationships between geographic information elements (i.e., topology) differentiates it from both CAM and the related but distinct automated mapping/facility management (AM/FM) systems [Kort]. As the reader may have guessed, a database is an important component of any GIS.

GIS can exist on any type of computer platform. PCs and workstations have become increasingly popular as GIS platforms over the past 10 years as their computational power has increased.

The development of computerized GIS goes back over 20 years. During the 1980s and 1990s GIS systems came to be widely used by local and national governments as well as by industry. Applications have included land use, environmental studies, census studies, and utility management. For telecommunication utilities, in particular, GIS provides a cost-effective way to capture the layout and location of geographically distributed facilities and users.

5.4.2 GIS in More Detail

The main components of a GIS are the following:

- tools for input and manipulation of GIS data
- database management system
- tools for geographic queries and analysis
- tools for geographic visualization

In the United States it is common to acquire map data by means of TIGER (Topologically Integrated Geographic Encoding and Referencing) files generated by the U.S. Geological Survey and Census Bureau. TIGER files contain information on U.S. transportation facilities (streets, highways etc.). Updates are produced every two years. A number of vendors produce enhanced TIGER files. A national topographic database was created in the United Kingdom during the 1980s. Similar efforts have been under way in other countries, including Japan and China [ESRI]. Maps of telecommunication/computing facilities (links, switches etc.) can be digitized either in-house or by outside contractors. Alternatively, map information can be scanned as a raster (bit-mapped) image to lower digitization costs, though this approach requires significantly more memory [Kort].

A file containing the physical boundaries of a geographical unit such as a suburban county will not occupy much computer memory. However if the file contains all the county's streets, highways, and railroads, it will occupy a considerable amount of memory. Files of 10 Mbyte or more are common. Thus, PCs and workstations in particular require powerful processors and ample memory to provide reasonable performance.

G. Korte reports that the database creation component of GIS accounts for approximately two-thirds of the cost of implementing a GIS [Kort]. Databases for GIS systems store two types of information: graphical (spatial) information relating to maps and **attributes** of the graphical map features. That is, attributes are descriptive data. Map data may include points [as an (x, y) coordinate], lines [as a series of (x, y) coordinates], and areas (as a series of boundary points). Certain GIS sytems store topological information that aids analysis:

Is a street one way or two way?

What is the shortest route from location A to location B?

Create an overlay of different map features.

Descriptive data is often stored in two-dimensional tables using the relational database model (see Section 5.3). Table 5.2 is a hypothetical attribute table. In general, there is a 1:1 relationship between features on a map and data records in the attribute tables [ESRI].

The inclusion of a database in a GIS system allows an environment for the use of tools for analyses of spatial data. For instance, one can query the GIS to determine how many customers are in a particular neighborhood or along a particular street. One can have the GIS display the route of a particular fiber optic run, the lines running through a particular street, or the combined facilities of two companies that are merging. One can also determine geographic areas for new facilities expansion by displaying the spatial distribution of customers not having SONET access lines.

Finally, a GIS provides tools for visualization. These include the ability to zoom in on specific subareas, the ability to display selected facilities, and the ability to create overlays of specific spatial and descriptive information.

Table 5.2 Example Attribute Table for a Relational Database

Feature #	Name	Location	Type
1	Router	HQ, Room 303	Data, Acme Model T
2	Switch	HQ, Room 44	ISDN, Beta Model F
3	Switch	HQ, Room 149	ISDN, Beta Model C
4	Remote switch	Bldg. K, Room 27	ISDN, Beta Model P
5	Fiber link	HQ, Room 149 to	Single mode
		Bldg. K, Room 27	

GIS systems have many advantages, once implemented. Among these are easy access to data, ease of information modification, ease of geographic analysis, increased productivity, lower costs, and the ability to standardize geographic information throughout an organization [Kort].

5.4.3 Conclusion

Geographic information systems can be seen as a specialized database technology, but also one with a wide variety of applications, including network planning.

5.5   EXPERT SYSTEMS

5.5.1 Introduction

An **expert system** is computer software that embodies and makes accessible the knowledge of human experts to provide computerized consultation on specialized knowledge. This consultation may take the form of troubleshooting, diagnosis, queries, prediction, or determining appropriate courses of action for given situations.

Expert systems have their roots in **artificial intelligence (AI)** research going back to the advent of computers after World War II. This research was originally intended to give machines the same sort of reasoning abilities people have. Early research attempted to do this in a general way that would be applicable to many fields. This effort was largely unsuccessful. Later it was realized that most expert knowledge is highly specialized. An expert's ability in facility location may be of little use in making traffic predictions, for instance.

This specialization of knowledge was taken into account in the first successful expert systems. Among them were DENDRAL (circa 1965–1970) for molecular structure identification, MYCIN from Stanford University (circa late 1960s) for blood infection diagnosis and treatment, PROSPECTOR from SRI (circa late 1960s–early 1970s) for geology, and XCON from Digital Equipment Corporation and Carnegie-Mellon University (circa 1980–1983) for computer system configuration [Jack 90] [Jack 92]. These were, in general, successful projects. MYCIN diagnosis and treatment compared favorably with that of expert physicians and was better than that of nonexpert doctors. It should be pointed out that MYCIN was never used in a real-world setting, though its descendants were. PROSPECTOR was able to find a major molybdenum deposit, and XCON was used by DEC to configure VAX[1] systems.

As computer power increased during the 1980s and 1990s, it became possible to implement expert systems in small platforms such as PCs and workstations. Durkin reports that over 60% of today's expert system applications

---

[1] VAX was a trademark of the Digital Equipment Corporation, which was acquired by Compaq in 1998.

are on PCs [Durk]. There is now widespread use of such systems by researchers and in industry and government.

5.5.2 Expert Systems in More Detail

What types of problem are amenable to expert system development? Generally, they must represent areas in which there is available human expertise. The availibility of the expertise is important because of the requirement for one or more experts (called **domain experts**) who can explain their reasoning to expert system developers and are available for a potentially lengthy development cycle. Moreover, the more specialized and limited in a well-defined way the expert's knowledge is, the greater are the chances of project success. It may not be possible to implement an expert system if large amounts of "commonsense" knowledge are called for. Also, people are very good at some activities (e.g., breathing, walking) but cannot explain how they carry them out.

There are five basic components to an expert system [Jack 92]:

- knowledge base
- inference engine
- knowledge acquisition system
- explanation system
- user interface (possibly graphical)

Knowledge acquisition involves transferring the expert knowledge to a computer. Usually in this phase an expert (or an expert team) works with a computer scientist knowledgeable in expert system technology. There has been a great deal of research effort on the topic of how "knowledge" should be represented in a machine. This is called **knowledge representation**. A common way of representing knowledge, known as **production rules**, has an IF ... THEN structure. For example, typical rules might be

IF a potential facility location is needed,

AND the site is not too expensive,

AND the site is centrally located,

AND a facility at the site will relieve congestion,

AND the facility budget is not exceeded,

THEN (recommendation): place a facility at the site.

Here is another example:

IF there is spare capacity in the cable duct,

AND it is the shortest or second shortest path from switch A to B,

AND the demand is less than the spare capacity the duct can carry,

AND the distance is short enough not to require repeaters,

THEN route the demand through the duct.

A typical expert sytsem may consist of hundreds or thousands of production rules. More sophisticated types of knowledge representation also are possible. These can use such concepts as logic, semantic networks, or frames [Jack 92].

A set of production rules reside in a **knowledge base**. An **inference engine** is the software that reasons through the rules to a conclusion.

In terms of the work of the inference engine, reasoning can proceed by way of either forward chaining or backward chaining. In **forward chaining** the system "reasons" from the data or facts to the conclusion. For instance, a transmitter failure may lead to the conclusion that the receiver will lose signal. In **backward chaining** the expert system reasons backward from the end result to causal conditions. This may be seen as a form of diagnosis or troubleshooting. For instance, from an optical signal loss an expert system may conclude that either the fiber optic cable was cut or the transmitter failed.

The **explanation system** conveys an explanation of the expert system's reasoning to the users. This is actually useful to both users and expert system developers, to be sure that the system reasons properly. Finally, a **user interface**, possibly graphical, provides ready access to the expert system.

5.5.3 Implementation

There are three main approaches to implementing an expert system: use a shell, use a programming language, or use an environment (toolkit) [Jack 92].

A **shell** is an expert system consisting of an inference engine and other systems, including knowledge representation but lacking a detailed knowledge base on a specific subject. One purchases the shell and adds the detailed knowledge base and reasoning. The first shell was EMYCIN, derived from MYCIN (the "E" stands for empty). Today, **frame-based development systems** are shells that combine production rules and object-oriented techniques [Durk].

Popular programming languages for expert systems (and artificial intelligence in general) include **LISP** and **PROLOG**. LISP (**List** Processing) utilizes the concept of lists. Its data structures and programs are lists. Moreover its basic operations are done on lists [Jack 90]. PROLOG (**Programming Logic**) is set up in a framework of objects and their relationships. As Jackson [Jack 92] puts it, a program in PROLOG is akin to a "database of facts and rules."

PROLOG was originally popular in Europe and LISP was originally popular in the United States. Both are artificial intelligence languages meant for symbolic processing. That is, they are specifically designed for problems of logic and knowledge. While computational languages such as C or C++ can be used to program expert systems, languages such as PROLOG and LISP, with their features for making inferences, heuristic search, and interaction, are better suited for this application.

A **toolkit** or **environment** consists of a language and frequently needed subroutines.

5.5.4 Conclusion

At present the biggest problem in expert system development is conveying the knowledge to machine form, rather than the software/hardware of the system. Most expert systems today are integrated with other software systems such as databases and management information systems [Lieb]. Expert systems have been widely used in finance, manufacturing, medicine, engineering, the sciences, and planning. As computer power continues to increase, it becomes even easier to deploy expert systems widely.

The relationship between expert systems and telecommunications is discussed in [Hedb]. Concise tutorials on expert systems can be found elsewhere [Durk] [Lieb]. Specific network planning implementations of expert systems also are available [Adam] [Ishi]. Expert system periodicals include *IEEE Expert: Intelligent Systems and Their Applications, Expert Systems and Their Applications, Intelligent Systems,* and the *International Journal of Applied Expert Systems.*

5.6   NONLINEAR MODEL OPTIMIZATION

5.6.1 Introduction

Models with continuously valued numerical variables often involve certain nonlinear dependencies. That is, the measurements may be a nonlinear function (known in statistics as a **regression function**) of the parameters. As an example, consider

$$y_i = \sum_{j=1}^{n} \theta_j f_j(t_i) \tag{5.1}$$

where y_i is a measurement at the ith time instant (t_i) and f_j is a (possibly) nonlinear basis function. Here, also, the n parameters, $\theta_j, j = 1, 2, \ldots, n$, appear linearly. Linear models, unless they are very large, are usually straightforward to solve. In a nonlinear model, however,

$$y_i = \Phi(\mathbf{\theta}, t_i) \tag{5.2}$$

Now, the relationships between the y_i and the θ_j (and t_i) may be nonlinear. In equation (5.2) $\boldsymbol{\theta}$ is the vector of parameter values and Φ is a nonlinear function. In a simple case the parameters may be real numbers.

Nonlinear model optimization starts with an **objective function**, $Z = \Phi(\boldsymbol{\theta})$. One must find the value of $\boldsymbol{\theta}$ that either minimizes (in a minimization problem) or maximizes (in a maximization problem) the value of Z. If Z is cost, one naturally wants to minimize it. In a curve-fitting problem the $\boldsymbol{\theta}$ may be parameters specifying different versions of a family of curves. The objective function would be a measure of fit (or alternatively, the error) of a parametrized curve to the data. In a large family of optimization problems, the $\boldsymbol{\theta}$ are real numbers.

As with the mathematical programs described in Chapter 2, an optimization problem may in general have a single, unique, **globally optimal** solution. It may also have many **locally optimal** solutions. A globally optimal solution is optimal (i.e., either minimal or maximal) with respect to the entire feasible solution space. On the other hand, a locally optimal solution is optimal with respect to only a limited region in the feasible solution space.

Figure 5.1 illustrates an objective function with a one-dimensional parameter space; there are two local minima and one global minimum. Naturally, with a two-dimensional parameter space the objective function surface would resemble hilly/valley-filled terrain if there were many local optima.

Descent techniques, such as the ones to be described below, basically start from some guessed point in the parameter space, estimate the slope, and move for the next computed guess in the downward (upward) direction if a minimization

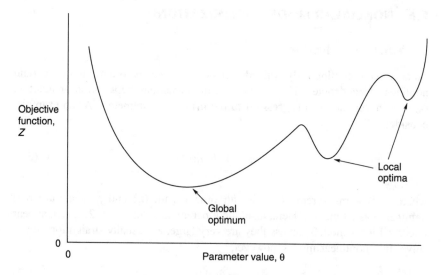

Figure 5.1 A typical objective function with a number of minima.

(maximization) is desired. Although pure descent strategies can converge to an "optimal" value very well, they can't tell the difference between a globally and locally optimal solution. Thus, if there are locally optimal solutions in the objective function, a descent technique is quite likely to converge to the closest locally optimal solution. Techniques to be discussed later such as tabu search, simulated annealing, and genetic algorithms, can aid a descent strategy in escaping from local optima.

5.6.2 Separable Models

Attempts to estimate such nonlinear parameters may exploit structure in the model. For instance, the following is an example of a "separable" model:

$$y_i = \sum_{j=1}^{n} \theta_j e^{-\alpha_j t_i} \tag{5.3}$$

Some parameters appear linearly (the θ), while others appear nonlinearly (the α). Numerical issues aside, the linear parameter estimates are unique functions of the nonlinear parametric estimates. That is, given the nonlinear parameter estimates, the linear parameter estimates can be "optimally" computed under a criterion such as least squares. This allows the iterative estimation to be conducted only among the space of nonlinear parameters [Dran] [Osbo]. This can significantly reduce the complexity of the optimization.

Examining the squared difference between the measurements and the model, or the "residual," one has

$$r(\theta, \alpha) = \sum_{i=1}^{m} \left(y_i - \sum_{j=1}^{n} \theta_j f_j(\alpha, t_i) \right)^2 \tag{5.4}$$

Basically, we wish to find the values of θ and α that yield the smallest value of the residual, $r(\theta, \alpha)$, error. One can think of this as "curve fitting." We wish to get a curve specified by $\sum_{j=1}^{n} \theta_j f_j(\alpha, t_i)$ to be as close to the measurements, y_i, as possible. Now

$$r(\theta, \alpha) = \|Y - \Phi(\alpha)\theta\|^2 \tag{5.5}$$

Here Y is the $m \times 1$ vector of measurements. Also $\Phi(\alpha)$ is the $m \times n$ matrix of function values. That is, the ijth entry is $f_j(\alpha, t_i)$. Finally, the notation $\|x\|^2$ is the notation for the sum of the squared values of the generic vector x.

$$r(\theta, \alpha) = \|Y - \Phi(\alpha)\Phi^+(\alpha)Y\|^2$$

Here $\Phi^+(\alpha)$ is the pseudo inverse of the matrix equation $Y = \Phi(\alpha)\theta$. This last equation contains the "variable projection" functional [Golu] [Osbo]:

$$r(\theta, \alpha) = \|(I - \Phi(\alpha)\Phi^+(\alpha))Y\|^2 \tag{5.6}$$

Here the parameters $\boldsymbol{\theta}$ do not explicitly appear. Now $\boldsymbol{\Phi}(\boldsymbol{\alpha})\boldsymbol{\Phi}^+(\boldsymbol{\alpha})$ is the orthogonal projector onto the linear space spanned by the columns of $\boldsymbol{\Phi}(\boldsymbol{\alpha})$ and $(\mathbf{I} - \boldsymbol{\Phi}(\boldsymbol{\alpha})\boldsymbol{\Phi}^+(\boldsymbol{\alpha}))$ is the projector onto the orthogonal complement of this space. Letting Pr be the projector, one has for the estimate $\boldsymbol{\theta}_j$:

$$\hat{\boldsymbol{\theta}} = \boldsymbol{\Phi}^+(\boldsymbol{\alpha})\mathbf{Y} \tag{5.7}$$

Here the $^\wedge$ indicates an estimate and:

$$r(\boldsymbol{\alpha}) = \|Pr_{\boldsymbol{\Phi}(\boldsymbol{\alpha})}^{\perp}\mathbf{Y}\|^2$$

One can go on to show that a critical point (i.e., minimum/maximum) in the reduced space is also one in the complete space. Although the variable projection technique reduces the dimensionality of the parameter space, optimization over a space of nonlinearly appearing variables is still necessary.

5.6.3 Descent Techniques

A number of iterative techniques are available for estimating parameter values that optimize objective functions. These optimization techniques iteratively develop an improved parameter estimate in each cycle based on the preceding estimate and on localized cost function derivatives. These methods perform these iterations based on a batch of data.

In the **Gauss–Newton** approach, N, measurement equations are modeled as

$$\mathbf{Y} = \boldsymbol{\Phi}(\boldsymbol{\theta}) + \mathbf{V}$$

Here $\boldsymbol{\Phi}(\ )$ is a function (often nonlinear) of $\boldsymbol{\theta}$, the parameter vector. Naturally, $\mathbf{V}$ is the measurement noise vector. Expanding the nonlinearity, one has

$$\mathbf{Y} = \boldsymbol{\Phi}(\boldsymbol{\theta})|_{\boldsymbol{\theta}=\hat{\boldsymbol{\theta}}_i} + \frac{d\boldsymbol{\Phi}(\boldsymbol{\theta})}{d\boldsymbol{\theta}}|_{\boldsymbol{\theta}=\hat{\boldsymbol{\theta}}_i}d\boldsymbol{\theta} + \mathbf{V} \tag{5.8}$$

where i is the current iteration. The least squares estimate of the deviation, $d\hat{\boldsymbol{\theta}}_i = \hat{\boldsymbol{\theta}}_{i+1} - \hat{\boldsymbol{\theta}}_i$, follows from

$$\min_{d\boldsymbol{\theta}} \|\mathbf{Y} - \boldsymbol{\Phi}(\boldsymbol{\theta})|_{\boldsymbol{\theta}=\hat{\boldsymbol{\theta}}_i} - \frac{d\boldsymbol{\Phi}(\boldsymbol{\theta})}{d\boldsymbol{\theta}}|_{\boldsymbol{\theta}=\hat{\boldsymbol{\theta}}_i}d\boldsymbol{\theta}\|^2 \tag{5.9}$$

or

$$\hat{\boldsymbol{\theta}}_{i+1} - \hat{\boldsymbol{\theta}}_i = d\hat{\boldsymbol{\theta}}_i = \left(\frac{d\boldsymbol{\Phi}^{\mathrm{T}}(\boldsymbol{\theta})}{d\boldsymbol{\theta}}\frac{d\boldsymbol{\Phi}(\boldsymbol{\theta})}{d\boldsymbol{\theta}}\right)^{-1}\frac{d\boldsymbol{\Phi}^{\mathrm{T}}(\boldsymbol{\theta})}{d\boldsymbol{\theta}}(\mathbf{Y} - \boldsymbol{\Phi}(\boldsymbol{\theta}))|_{\boldsymbol{\theta}=\hat{\boldsymbol{\theta}}_i} \tag{5.10}$$

It can be seen that the Gauss–Newton approach involves a linearization of the nonlinearity. More specifically, it involves computing the least squares estimate of the correction term, $d\hat{\boldsymbol{\theta}}$.

For the **Newton** method, sometimes referred to as the **steepest descent** method, we write

$$\hat{\boldsymbol{\theta}}_{i+1} - \hat{\boldsymbol{\theta}}_i = d\hat{\boldsymbol{\theta}}_i = \frac{d\boldsymbol{\Phi}^{\mathrm{T}}(\boldsymbol{\theta})}{d\boldsymbol{\theta}}(\mathbf{Y} - \boldsymbol{\Phi}(\boldsymbol{\theta}))|_{\boldsymbol{\theta}=\hat{\boldsymbol{\theta}}_i} \tag{5.11}$$

Here we are proceeding in the direction of steepest descent down an error (objective function) surface.

It has been empirically observed [Marq] that the steepest descent algorithm often converges slowly to the optimal parameters estimate. That is, initially fast convergence changes to slow convergence. By way of constrast, the Gauss–Newton approach is most effective when the estimates are close to the optimal solution. Naturally, the Taylor series on which the Gauss–Newton method is based is most accurate when the estimate is close to the optimal parameter values.

A method based on the strong features of both techniques is due to Marquardt [Marq]. In the Gauss–Newton approach,

$$\mathbf{P}d\hat{\boldsymbol{\theta}}_i = \mathbf{g} \tag{5.12}$$

where the following equations must be solved:

$$\mathbf{P}^{-1} = \left(\frac{d\boldsymbol{\Phi}^{\mathrm{T}}(\boldsymbol{\theta})}{d\boldsymbol{\theta}}\frac{d\boldsymbol{\Phi}(\boldsymbol{\theta})}{d\boldsymbol{\theta}}\right)^{-1}|_{\boldsymbol{\theta}=\hat{\boldsymbol{\theta}}_i} \tag{5.13}$$

$$\mathbf{g} = \frac{d\boldsymbol{\Phi}^{\mathrm{T}}(\boldsymbol{\theta})}{d\boldsymbol{\theta}}(\mathbf{Y} - \boldsymbol{\Phi}(\boldsymbol{\theta}))|_{\boldsymbol{\theta}=\hat{\boldsymbol{\theta}}_i} \tag{5.14}$$

Now consider

$$(\mathbf{P} + \lambda\mathbf{I})d\hat{\boldsymbol{\theta}}_i = \mathbf{g} \tag{5.15}$$

If λ is small, **Marquardt's** approach operates approximately as a Gauss–Newton approach. If λ is large, an approximate Newton method results. The parameter λ can be adaptively set so that the optimization is initially Newton-like and later is Gauss–Newton-like near the optimal parameter estimates (Figure 5.2). Various heuristic adaptive programs for λ are possible.

5.7   TABU SEARCH

5.7.1 Introduction

As has been mentioned, pure descent strategies will converge to a local optimum that is nearby in the parameter space, rather than a global optimum that is farther away. This problem has long been recognized in optimization. A simple way to improve one's chances of finding a global optimum is to run a descent technique from many randomly chosen initial starting points in the parameter space. While

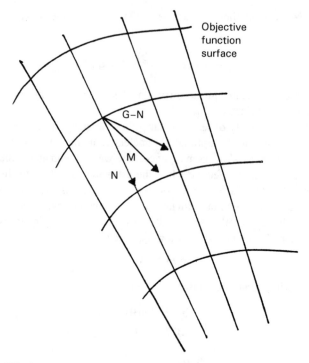

Figure 5.2 Objective function surface descent: N, Newton method or steepest descent; G–N, Gauss–Newton method; M, Marquardt Method.

this may work in low dimensional problems with few and well-behaved minima (maxima), it is inefficient for the complex high-dimensional problems that people would really like to solve today.

Tabu search is a heuristic approach used to direct techniques, such as descent algorithms, to find a good minima/maxima even in the presence of local optima [Glov 86] [Glov 90] [Glov 93a] [Glov 93b]. Though elements of tabu search were used as early as the 1960s, it was F. Glover in the mid-1980s who helped to set up this approach as a distinct area of study. Since then there has been a good deal of work on the subject by a variety of researchers, starting with those in scheduling.

The key aspect of tabu search is that it incorporates memory regarding the optimization being performed to make better choices of parameters. To ensure that overall optimization algorithm will be able to escape from local optima certain types of movement in the parameter space are made "tabu" (taboo). If certain "aspiration" criteria are met, a tabu move may no longer be tabu. One can also use memory functions of different durations to provide diversity and intensity to the search.

What kinds of problem are amenable to tabu search? Generally, they are optimization problems in which a "move" can "transform" one solution state into another and there is an objective function for determining the value of moves. A move could involve the value of a parameter estimate, as in Section 5.6, inserting and removing items in a set, interchanging the positions of packets in a buffer, or changing the order of cities visited in a traveling salesman problem.

Tabu search has been applied to a wide variety of optimization problems with a good deal of success. It can find excellent solutions and is relatively intuitive and simple to implement; and it is relatively straightforward to incorporate new forms of constraints to existing optimization problems. Tabu Search has been applied to such generic problems as scheduling, graph problems, sequencing, traveling salesman problems, and assignment [Glov 90]. Network planning examples include topological design [Lee 89a], and path assignment in tele-communications networks [Ande 93] [Oliv]. Tabu search can be integrated with such other techniques as neural networks, branch and bound, cutting plane techniques and simulated annealing [Glov 90] [Glov 93a]. Tabu search can and should be tailored—usually in a straightforward manner—to individual optimization problems.

5.7.2 Tabu Search in More Detail

Glover has stated [Glov 90] that tabu search is founded on three main "themes." They are as follows.

1. The use of flexible attribute-based memory structures, through which better use can be made of objective functions and historical search information than is possible in rigid memory structures (e.g., branch and bound).

2. A control based on the interaction between tabu restrictions and aspiration criteria. This control make use of the memory structure.

3. The inclusion of memory functions with different time durations (short term to long term). One purpose of this feature is to allow **intensification** of the search. Intensification strategies produce a bias toward move types and solution charcteristics that have a historical basis as being good. Another purpose is **diversification** of the search. Diversification produces a bias to send the search into new regions.

What is the philosophy behind **tabu restrictions**? The idea is to prevent the reversal/repetition of certain types of move with specific attributes, and thus to avoid "cycling" that returns an optimization procedure to a local optima. Put another way, the use of restrictions allows an optimization procedure to escape from local optima while still making valuable moves.

Generally at each move a list of potential candidate moves is created. These are then evaluated, either directly (if the associated computation can be done efficiently) or approximately (if the computation is complex or not possible at that move).

In the early history of tabu search the most effective tabu list sizes tended to fall in the interval from 5 to 12, with 7 being quite efficacious. There was thus some speculation about a relationship between this and the similar number of items usually present in human short-term memory. Could evolution have resulted in a short-term memory of about 7 items as good for problem solving? Later, optimization problems were found in which the best tabu list sizes were related to the dimension of the problem. Glover has suggested that people may naturally select attributes that are effective for the size of their short-term memories. Since humans picked the attributes for the early tabu restrictions, based on intuition, this may account for the similar sized values [Glov 90].

Aspiration criteria allow suspension of the tabu restrictions if a significant improvement in the solution can be obtained. Aspiration criteria can be implemented on either short-term basis or a long-term basis. There are subtle interactions between tabu restrictions and aspiration criteria that must be considered.

In **target analysis** some initial effort is made to find representative optimal or nearby optimal solutions for a class of problems. These solutions, known as target solutions, can then be used to evaluate potential moves that will be likely to be successful/unsuccessful. Such analysis can also help to develop intensification and diversity strategies. In fact, the search can be aided by subjecting moves that are likely to be successful to prior statistical analysis.

It should be mentioned that diversification is particularly needed when one wishes to cross certain objective function hills/valleys to move toward a good solution. Again, it is possible to guide the diversification by gathering statistics from a number of earlier iterations to categorize the benefits of moves within a distance class. A statistical analysis can also be used to favor starting points for an optimization that are significantly different from ones chosen earlier.

5.7.3 Conclusion

The creation and the continued development of tabu search represent an effort to introduce artificial intelligence techniques into a field, optimization, formerly characterized by mathematically well-understood techniques that were not always completely effective. While tabu search has certainly proven to be effective, its implementation has outpaced our analytical understanding of its operation and convergence properties. Actually this is true of other software-related fields and is one of the things that makes tabu search so interesting.

5.8   SIMULATED ANNEALING

5.8.1 INTRODUCTION

Tabu search is not the only technique available to perform an optimization in such a way that the algorithm can escape local minima/maxima. Another such technique is **simulated annealing**. It is also applicable to combinatorial problems, where one solution can be transformed into another by a "move" [Otte] and there is an objective function available for evaluating the quality (value) of a solution.

A commonly used example of a combinatorial problem is the **traveling salesman problem (TSP)**. The goal in this problem is to pick the best route to visit each of N cities once and then return to the starting city. "Best" here means the route involving the shortest distance traveled. The solution can be viewed as a vector in which each city is listed once, in the order in which it is visited.

A brute force enumerative solution simply won't work for a combinatorial problem like the TSP. A little thought will show that if N is the number of cities, then there are $(N - 1)!$ solutions. For $N = 50$ cities, the number of possible solutions is 6.08×10^{62}. Suppose we had a million processors, each capable of evaluating 100×10^9 solutions a second (characteristics beyond technology today). Then it would take 1.92×10^{38} years to evaluate all possible solutions. This number is substantially greater than the number of years in the age of the universe! In fact, the traveling salesman problem belongs to a large group of related computational problems, known as being NP complete, for which no fast optimal algorithm is expected to be found.

Yet there are very practical reasons for wanting to solve this problem. Of course in real life a salesman doesn't set off to visit 50 cities in one trip. But analogs of the traveling salesman problem arise in real problems. For small N the traveling salesman problem is a good model for deliveries by vehicles and service personnel routing. For large N, S. Carlson [Carl] describes a particularly interesting application that required him to figure out the best sequence of galaxies a telescope should point to in a night's automated viewing to minimize the movement of a 40-year-old drive mechanism. Another application is the drilling of holes in circuit boards. An automated drilling system moves the board under the drill by the precise increments necessary to ensure that each hole in the series is drilled at the right spot. The solution of the associated traveling salesman problem goes a long way toward minimizing the amount of time needed for drilling the holes. Related problems occur for VLSI placement for modules and routing.

There is, in fact, a large family of practical combinatorial optimization problems like the traveling salesman problem, which are impossible to solve by brute force enumeration. Many of these problems are relevant to network planning. And many of these problems can be solved by the algorithm known as simulated annealing.

5.8.2 The Simulated Annealing Algorithm

If we relax the constraint that we want a globally optimal solution for every potential problem instance, there are many heuristic algorithms available that will usually find a solution with an objective function value within a few percent of the optimal solution and sometimes will even find a globally optimal solution. These include tabu search, genetic algorithms (see Section 5.9), and simulated annealing. Such solutions will be fine for the practical problems mentioned, including those in network planning.

How does simulated annealing work? One can consider an algorithm where, instead of always choosing the next solution to be one with an improved objective function, as in a pure descent technique, one instead occasionally (probabilistically) chooses the next solution to be one with a worse objective function. The real question is how one makes this choice.

N. Metropolis and a number of coworkers [Metr] found such a good choice based on intuition regarding statistical mechanics. In the "Metropolis loop" of a simulated annealing algorithm, one calculates the *change* in objective function in making a particular move. A move that worsens the cost function is accepted with probability:

$$\text{Prob}_{\text{Accept}} = e^{-\Delta E / kT} \tag{5.16}$$

where e is the irrational number ($= 2.71828\ldots$) that can be found on any scientific calculator. Then ΔE is the change in objective function (energy according to the intuition), k is a (Boltzmann) constant relationship between temperature and energy, and T is "temperature".

In simulated annealing the "temperature," T, is initially set at a high value so that the probability of acceptance is close to one and is slowly reduced until the probability of acceptance is close to zero. Thus, initially there is a high probability of accepting worse solutions, and eventually only improved solutions are accepted.

In a statistical mechanics context this procedure leads to a distribution of states according to a Boltzmann probability density:

$$\frac{e^{-E_i/kT}}{\sum_j e^{-E_j/kT}} \tag{5.17}$$

where E_i is the energy of the ith state. The denominator is included for the purpose of normalization (so that the overall quantity is a probability between 0.0 and 1.0).

Simulated annealing "simulates" the annealing process that takes place in certain materials. If the material is melted at a high temperature and then slowly, cooled, a macroscopic ordered crystalline structure can result. This is also a low energy system state. The proponents of simulated annealing feel that there is a connection between this annealing process and combinatorial optimization problems.

5.8.3 Simulated Annealing in More Detail

As Otten and van Ginneken point out [Otte], at high temperatures each state is (approximately) equally likely to be the current state. At low temperatures only low energy states have a significant probability of occurring.

Although the original insight into the relationship between the statistical mechanics of annealing and optimization goes back to 1953 [Metr] it was not until the 1980s that this insight was fully exploited [Čern] [Kirk].

Practioners of simulated annealing find it relatively straightforward to apply to various combinatorial problems. Some tailoring to each problem is necessary, though. Care must be taken in selecting moves. Depending on the particular problem, some moves are better than others, and naive ones may be terrible. It also is beneficial to avoid moves that most probably will be rejected [Otte].

The choice of the temperature schedule is also important. If the temperature decreases too rapidly, a low energy (nearly optimal or optimal) solution may not be reached. If the temperature decreases too slowly, simulated annealing may in fact be slower in convergence than other possible algorithms. Adaptive temperature schedules are possible [Otte].

Speedup techniques have also been considered for simulated annealing. In a two-stage version one first finds a good (but not by any means nearly optimal) solution through heuristic techniques and then makes use of simulated annealing. In some problems, when the objective function is computationally expensive, moves may be evaluated approximately.

Descriptions of work using simulated annealing for network planning problems at British Telecom have been published [Crab 94] [Crab 95] [Fish].

5.9   GENETIC ALGORITHMS

5.9.1 Introduction

In Section 5.8 we saw how a certain insight from physical metallurgy was used as the basis of the simulated annealing optimization algorithm. In fact, such a use of knowledge from the sciences is not an isolated case. A great deal of effort has taken place in recent years in applying concepts from evolutionary biology to optimization. Perhaps the most successful example of this approach is called

genetic algorithms. Genetic algorithms belong to a larger family of algorithms known as evolutionary algorithms [Mich] [Sanc]. They are efforts to apply concepts from the theory of biological evolution, such as natural selection, reproduction, genetic diversity and propagation, species competition/cooperation, and mutation, to search, optimization, and machine learning problems.

In a genetic algorithm an "individual" in a population represents a particular solution to the problem at hand. Individuals may be represented by binary strings or other representations, depending on the problem. Over a number of generations of individuals, it is seen that "the fittest survive," with the phenomenon taking place at electronic speeds, and the best solutions present in the final population.

The concept of genetic algorithms began with the work of J. H. Holland [Holl]. Genetic algorithms have been applied to a variety of optimization problems, including those of the combinatorial type [Davi] [Gold 89]. They are certainly applicable to specific network planning problems [Paul].

5.9.2 Genetic Algorithms in More Detail

Like a natural population, a population of solutions in the genetic algorithm approach reproduces in such a way that useful characteristics are reinforced in the population and harmful characteristics are eliminated. As Tomassini [Toma] points out, nature is the inspiration for genetic algorithms, but one can design the algorithms in a pragmatic way that does not follow nature "to the hilt."

Populations in genetic algorithms are generally of a constant size with N individuals. Each individual is just a string of (possibly binary) symbols. Some method is available for evaluating the "fitness" (really the objective function) of each individual. Let f_i be the ith individual's fitness. For a new population, individuals are often chosen in proportion to their relative fitness, p_i, with probability:

$$p_i = \frac{f_i}{\sum_{i=1}^{N} f_i} \tag{5.18}$$

The effect of this choice is to bias the population toward containing more individuals that are "fitter." Naturally, though, something is needed beyond this random selection to create new individuals. Two population mechanisms for introducing new individuals are **mutation** and **crossover**. In mutation, the more limited mechanism, bits in an individual are changed at random with a very small independent probability. This can be seen as an attempt to prevent convergence to a local minimum/maximum by sampling other points in the solution space. In the more powerful crossover, characteristics of two parent individuals are combined to form two new individuals according to one or more rules that make sense for the problem at hand. An individual will take part in a crossover with some independent probability, which is often about 0.6 [Toma].

Not everyone agrees that crossover is more powerful than mutation or even that these mechanisms should be used together. Some other evolutionary algorithms use only selection and mutation [Foge].

It has been mentioned that binary codes are often used to represent solutions. This is not always the best choice. For example, numbers that are numerically close together may "look" quite different (i.e., have a large Hamming distance) when expressed in binary form. Gray code can be used to mitigate this problem. In a Gray code consecutive numbers differ by only one bit. For combinatorial problems in particular (the traveling salesman problem, partitioning, etc.), the choice of a good representation depends on the problem, and a good choice is not always the most obvious one.

More sophisticated selection procedures, are also possible [Gold 89] [Toma]. These attempt to overcome problems of fitness–proportionate selection type just discussed. One problem is that as a genetic algorithm proceeds and as the population comes to consist largely of fit individuals, the differences in fitness becomes small, and the use of fitness–proportionate selection is no longer justifiable or efficient. A second problem is posed in the presence of a super-individual having a very high fitness. Unless preventive steps are taken, this individual's characteristics will spread throughout the population quickly.

Other ideas have been borrowed from the theory of natural evolution to make better genetic algorithms. For instance, if one is interested in determining multiple optima, one would like subpopulations to cluster around each (local or global) optimum. This is akin to **niching** in biology. Related to this is **coevolution**, which occurs when different species compete or cooperate with one another. Such an approach has received attention during the 1990s. Hillis [Hill 93], for instance, used two species to develop a sorting networks. One of the species represented sorting networks and the other species represented sorting problems. The first species was evolved to do sorting well in cooperation with the second species, which produced difficult sorting problems for the first species.

Usually the whole population is replaced after each cycle. This is referred to as a **generational** approach. Other possibilities have certain advantages. For instance, one could delete only the worst individuals from each cycle.

5.9.3 Parallel Algorithms

The development of high performance parallel machines opens new possibilities for genetic and evolutionary algorithms. One advantage of the use of parallel processing is simply raw speed. Expensive functions, such as fitness calculations, can be spread among multiple machines to provide faster solutions. Alternatively, since a genetic algorithm, is a stochastic algorithm, multiple independent programs may be run on different machines and the best of the solutions chosen as the result.

An even more interesting use of parallelism is to mimic the spatial distribution of natural populations. One may have an **island** model with distinct population centers and loose coupling in the form of a limited number of "visitors" from population center to population center. This model is suitable for a **multiple instruction, multiple data** stream machine (MIMD) or for a network of workstations. In a **fine-grained** (or grid or cellular) approach, one individual is placed on each node of an interconnection network (e.g., a two-dimensional grid or a toroidal grid). Selection and reproduction operations occur locally [Toma]. This is suitable for a **single instruction multiple data** stream (SIMD) machine.

5.9.4 Conclusion

It is reemphasized that genetic algorithms are part of a larger family of evolutionary algorithms. Among these are **genetic programming** algorithms. Here short computer algorithms (usually in C or C++), rather than symbol strings, represent each individual. One can also have **hybrid genetic algorithms**, in which the genetic approach is married to specialized optimization procedures for specific problems.

Like tabu search and simulated annealing, our mathematical understanding of genetic algorithms is lagging behind our implementation abilities.

The reader interested in genetic algorithms can find more information in books [Foge] [Sanc] [Toma] and in the journals *IEEE Transactions on Evolutionary Computation* and *Evolutionary Computation* (from MIT Press Journals).

5.10   NEURAL NETWORKS

5.10.1 Introduction

Over the past 15 years a great deal of progress has been made on a computation model radically different from the stored program control of the more traditional von Neumann model that has dominated machine computation since World War II. This model is known as the **artificial neural networks** model, or simply neural networks. Artificial neural networks are machine implementations (either hardware or software) of brain-inspired circuitry that are well suited for certain types of pattern-matching problems.

It should be mentioned early on that while there is no telling what the future may hold [Gari], at present machine implementations are limited in their abilities compared with animal brains. To get some idea of the relative complexities involved, let's start with the human brain, which has about 100 billion brain cells known as **neurons**. Each neuron has connections to about 10,000 other neurons. Thus there are about 10^{15} **synapses** (connections).

Roughly speaking, when a particular neuron has received enough signal strength from its "inputs" from other neurons, the receiving neuron releases a signal burst over its output connections to other neurons. Artificial neural networks minic this behavior, though not exactly. Today's artificial neural networks consist of a few thousand neurons at most. Simulations of neural networks on a PC or workstation are comparable in complexity to the brain of a leech. Artificial neural networks are approaching the complexity of a fly or bee. The human brain is about eight orders of magnitude more complex than today's artificial neural networks [Gles].

One may ask why scientists and engineers would be interested in building systems no more complex than those of insects. One may also ask why they would be interested in mimicking natural thought processes when a standard computer can execute millions of instructions a second. The answer in both cases is that biological systems are far better at pattern recognition type problems than today's best von Neumann model based computers operating under conventional programming. The natural brain's ability to distinguish individuals (in terms of faces, for instance), recognize handwriting and speech, to follow a single conversation in a roomful of talking people, and to fill in parts of missing (and noisy) patterns is unsurpassed. These are also problems that are difficult to solve on today's conventional computers. By way of contrast, neural networks are ill suited for detailed numerical computations that von Neumann machines excel at.

Thus neural networks offer a possible approach for network planning problems such as fault location [Tind], traffic prediction, and facility location.

5.10.2 Neural Networks in More Detail

Artificial neural networks gain their expertise on specific problems through learning by example rather than by detailed programming or programmed rules. For a neural network doing character recognition, for instance, the neural network is "shown" written examples of letters and is "told, in a **training session**," which letters they are. Training sessions can be lengthy. But after a sufficiently well-designed training session, the neural network will be able to recognize handwritten letters it has not seen before with a surprisingly high probability of being correct. If this seems like magic, remember that people and animals have been "learning" for millennia, and that our ability to implement neural networks is outpacing our ability to understand their detailed behavior.

But how are artificial neural networks implemented? In a typical backpropagation neural network, the "neurons" may be placed in three layers: an input layer, one or two hidden layers, and an output layer. Signals enter the input layer neurons in parallel. There are connections from the input layer neurons to the hidden layer neurons. There are also connections from the hidden layer neurons to the output layer neurons. The (parallel) outputs are then taken off the output layer.

The reader with some experience with networks will recognize what is being described as a **feedforward network**. Many neural network architectures have been studied, but the two major categories are feedforward and **feedback or recurrent networks**. In a feedback network connections may go backward from the outputs of one layer to the inputs of the preceding layer or the same layer.

Connections between neurons may be **inhibitory**, tending to prevent neuron firing, or the connections may be **excitatory**, tending to promote neuron firing. In a basic neural network, the output of a neuron may be a weighted sum of its inputs [Stan]:

$$O_j = \sum_{i=1}^{N} w_{ij} O_i \qquad (5.19)$$

where O_i is the numerical output of the ith neuron and w_{ij} is the numerical weight on the connection from the ith neuron output to the jth neuron input. Excitatory weights can be positive and inhibitory weights can be negative. Sometimes the w_{ij} and O_i values are constrained to be integers. Moreover, the "transfer"-like function of equation (5.19) is linear, but nonlinear functions can also be used.

During training using **supervised learning** each output neuron is "taught" what its response to training samples should ideally be. In the different case of **unsupervised learning**, an outside source is not used to correct the network.

One type of feedback network is the **Hopfield network**. Each neuron output can be either $+1$ (on) or -1 (off). In a Hopfield network each neuron is connected to every other neuron. While popular in the later 1980s, Hopfield networks today are mainly of historical interest [Gles].

A popular form of feedforward network relies on **back propagation**, which in turn relies on supervised learning with continuously valued variables. Back propagation was discovered as early as the late 1960s but did not come into widespread use until the mid-1980s. During training on input/output pairs, local errors are propagated in the reverse direction (backward).

Finally, **Boltzmann machine** architectures have some interesting properties. Connections are bidirectional. There are input, hidden, and output neurons. Neurons can be arbitrarily connected (not necessarily in a layered model). Neuron states are also binary (either 0 or 1). A simulated annealing-like procedure, complete with energy function and temperature, is used to randomly "activate" (set to one) states.

One nice property of neural networks is reliability. If a single bit is flipped or a single transistor shorts in a von Neumann machine, it is likely that the computer/program will not function. Experiments with neural networks have demonstrated that, in general, a small number of random changes in weights or connections leads to a **graceful degradation** in the neural network's performance. That is, the network continues to work, though with somewhat less accuracy. This phenomenon is due to the distributed nature of information in a neural network.

There are a number of ways to implement a neural network. The simplest and most flexible, but lowest performance alternative, is to implement the neural network in the form of a program on a conventional PC or workstation. That is, neuron states can be stored in conventional memory and updates can be made using stored connection weight patterns. Moving in order of increasing performance, other implementations are general-purpose digital neural computers, special-purpose digital neural hardware, and analog chips. All these approaches are order-of-magnitude limited to about a million synapses (connections). Optical hardware can improve on this by a factor of 10 to 100, but at this time optical implementations exist only in the laboratory.

5.10.3 Conclusion

Artificial neural networks are useful because of their abilities to learn (absorbing patterns from data), to generalize (process incomplete or noisy information), to be reliable, and to perform functions essential to biological organisms but beyond the routine capabilities of conventional computers to date.

Some suggest [Gari] that we are reaching a technological mismatch between our capacity to design systems and our capacity to implement systems (VLSI in particular). That is, our design abilities are lagging behind our implementation abilities. Artificial neural networks may be an approach that fills this gap. The reader interested in artificial neural networks can find useful material in several fine publications [Bras] [Gles] [Simp] [Stan].

Good sources for information on neural networks are the journals *IEEE Transactions on Neural Networks*, and *Neural Computation* and *Neural Networks* from the International Neural Networks Society. A discussion of both neural networks and evolutionary computing was published in 1996 [Tsui].

5.11 PROBLEMS

1 What is object-oriented programming?

2 What is an object?

3 What are some advantages of object-oriented programming?

4 What is the object-oriented concept of encapsulation?

5 What is the object-oriented concept of inheritance?

6 What are the five common database models?

7 When was the relational database model invented?

8 Create a hypothetical relational table for five routers for a company based at two sites.

9 Discuss some of the advantages and disadvantages of using a mainframe as a platform for a database.

10 Discuss some of the advantages and disadvantages of a client/server system architecture for a database.

11 What is a geographic information system (GIS)?

12 What are the major components of a GIS?

13 What is a TIGER file?

14 How are map features stored in a GIS?

15 What is the significance of incorporating a database of feature attributes in a GIS?

16 Discuss the original goal of artificial intelligence (AI) research. How does work on expert systems differ from this goal?

17 What are the basic components of an expert system?

18 Create a hypothetical production rule.

19 What is the difference between backward chaining an forward chaining?

20 Discuss the differences between LISP and PROLOG.

21 What is the difference between a globally and a locally optimal solution? Why do locally optimal solutions cause problems for optimization techniques?

22 Intuitively, how does a descent technique work?

23 Create a hypothetical separable model.

24 How is the Gauss–Newton approach to function optimization related to least squares estimation?

25 How is the parameter λ used in the Marquardt approach to function optimization?

26 Where does the "tabu" in tabu search come from?

27 What types of optimisation problem are well suited for tabu search?

28 What are tabu restrictions and aspiration criteria?

29 Discuss Glover's three main themes for tabu search.

30 Discuss the apparent relationship between human short-term memory and the size of tabu lists in early research on tabu search.

31 What physical process does simulated annealing mimic?

32 Explain the details of the simulated annealing algorithm.

33 Discuss the choice of the temperature schedule in simulated annealing.

34 When was the original work on simulated annealing performed? When did it fully develop?

35 Explain why the traveling salesman problem cannot be solved through brute force enumerative solution. Use $N = 75$ cities as an example.

36 What is the approach behind genetic algorithms in particular and evolutionary algorithms in general?

37 Discuss the concept of fitness in a population and its use in genetic algorithms.

38 Discuss mutation and crossover.

39 How are "individuals" represented in genetic algorithms?

40 How can parallel processing be used in genetic algorithms?

41 What type of application are neural networks well suited for?

42 How is a neural network "programmed"?

43 What are the two major categories of neural networks? Given an example of each.

44 Discuss the reliability of neural networks.

45 Discuss the implementation of neural networks.

46 Computer Project Solve an optimization problem using either nonlinear function minimization, tabu search, simulated annealing, genetic algorithms, or neural networks. Consult the instructor for the exact problem(s) to be solved. Hand in a report consisting of a problem and solution approach description, flowchart, results (graphs are recommended), and a code listing.

6

DATA ANALYSIS FOR PLANNING

6.1 INTRODUCTION

In planning networks it is sometimes necessary to analyze various quantities. These include traffic statistics or projections, measure of performance, or measures of quality of service. Often records of past data must be summarized in a few well-chosen statistics. Sometimes it is necessary to extrapolate or predict future performance from these records.

This chapter presents an introductory look at data analysis, including the modern areas of network visualization and data mining. The section on statistics useful for data mining discusses regression, discriminant analysis, logistic regression, cluster analysis, time series, simulation, and rule generation. This material is followed by an extensive analytical introduction to least squares estimation, the most widely used theory of statistical estimation. The chapter concludes with a discussion of Kalman filtering—a very flexible twentieth-century tool for estimation that has been used for network planning.

6.2 NETWORK VISUALIZATION

6.2.1 Introduction

Operators of networks, be they common carriers, companies, governments, or other organizations, find that making sense of network data is important but not straightforward. In networks of up to hundreds of switches and thousands of links, "visualizing" the current state or a future state of the network can be demanding. Our discussion on this topic relies largely on the work of R. A. Becker, S. G. Eick, and A. R. Wilks of AT&T Bell Laboratories [Beck 91] [Beck 95] [Eick].

Questions a network planner would like to answer include the following [Beck 95]:

- What geographic areas have the most over- (under-)utilized links?
- Which links/nodes are the most over- (under-)utilized?
- What is the effect of adding capacity in certain locations?
- What is the effect of a particular rerouting?

Network visualization is distinct from the usual scientific visualization problems because three-dimensional physical objects are lacking and because of the uniqueness of the graph-based representation used in network visualization. Moreover, the larger the network, the more important it is *not* to represent microscopic detail (e.g., individual calls or packets) and to *surely* represent macroscopic quantities (e.g., link/node utilization).

6.2.2 Network Visualization in More Detail

A number of typical attributes and features can be displayed [Beck 95] [Eick]. One is statistics. There is a need to display both absolute and relative (e.g., percentages) values. Also useful are transformations of data such as square roots and logarithms. Sometimes it is valuable to display statistics that are within certain numerical ranges, to alleviate clutter and ensure a focus on relevant information. Thus one might choose to display only switches with utilizations above 80%. Users also need the capability to aggregate statistics in selected regions.

Geographical displays are also of interest. As with geographic information systems, being able to zoom in to a particular subarea is particularly useful. Certain nodes/links also may be deactivated to accentuate the display of the remaining parts of a network.

Colors can be used to accentuate particular information. For instance, red may indicate overloaded links. Finally, time animation is beneficial for displaying network evolution over a certain time period.

There are four main approaches to displaying network data [Beck 95] [Eick]. The first uses **link maps**; that is, links are drawn between nodes with link parameters displayed numerically, in the line color or by line thickness. While intuitively appealing, this approach has some inherent difficulties. For instance, long links between perimeter points may cover and obscure information in the middle of the display. Drawing important links (e.g., those with high utilization) last is one way to prevent them from being obscured. Displaying only some links (possibly using threshold information) is another possibility.

Link information for large carriers can be clearer if displayed on a **3D global display**. Here the routes circle the sphere of the earth, much like airline routes [Eick]. This technique allows easy orientation and enhanced clarity.

Network data representations such as link maps are sometimes plotted without regard for geography, as in traffic on a single local area network. Nodes are often placed in a circle, to highlight high traffic pairs [Farr] [Sevi]. Heuristic algorithms can be used to place the nodes on a display so that node pairs with a high degree of interaction are physically close on the display.

In a **node map**, only statistics associated with nodes are displayed. One or more statistics can be displayed in the node symbol itself by means of color, shape, or size. Unlike link maps, node maps do not have the problem of certain information hiding other information.

Finally, **matrix displays** of information are possible. Here information is displayed in a (mathematical) matrix format with the ijth element representing link traffic between nodes i and j. The ijth element itself is a small (possibly colored) symbol such as a square or circle. Overlooked nodes may show up as red rows or columns. Naturally, one can easily lose a sense of geography in such a format. The numbering of rows and columns in particular has a strong influence on a user's ability to make sense of the results.

6.2.3 Conclusion

A picture is worth a thousand words. Proper network visualization tools can be very useful to the network planner. Certain early basic work in network visualization was done by Bertin [Bert 81]. The work of Becker, Eick, and Wilks is written up largely in the context of network management, though many of the concepts are generic and easily carry over to network planning. Many of the features discussed above appear in the (AT&T) SeeNet package. Network monitoring issues related to network visualization are discussed in [Farr]. A network planning visualization tool developed at NYNEX for interoffice network planning is described in [Sevi].

6.3 DATA MINING

6.3.1 Introduction

Data analysis has been around as long as the development of science and business. However, until the mid-twentieth century it was largely a manual exercise. Large amounts of data could be collected, but a major data analysis limitation was the immensity of the human effort needed to sift through the data. Two converging technologies have now made possible a breakthrough in data analysis. One is the continued development of databases, be they small or large (see Section 5.3 for more on databases). A second technology, with roots in the artificial intelligence field, is the automated discovery of "knowledge" (i.e., rules, patterns, trends, associations) in data by computers. When such automated tools are used in conjunction with modern databases, the resulting search capabilities are known today as **data mining**.

Data mining is seen today as an up-and-coming technology for businesses seeking to gain competitive advantages from their data stores. Data mining also has important applications in science, government, and law.

Data mining addresses an important technological mismatch in computer technology. Computers are capable of generating far more information than we are capable of analyzing. Even 25 years ago it was fascinating to watch a line printer spew out page after page of information. Today, for instance, earth resource satellites generate far more data than can be possibly examined manually. In business, the widespread automation of records and transactions has led to the accumulation of large amounts of data, necessitating a new activity called **data warehousing**.

As R. Barquin, president of the Data Warehousing Institute points out [Barq], most *Fortune* 1000 companies have now initiated data warehousing efforts. To give some feel for the amount of data involved, note that MasterCard International's data warehouse, created in 1995, held a terabyte of data (spread over 350 disk drives!) [Free]. Communication companies are no exception to this trend. MCI has a 2.5-terabyte data warehouse [Nadi].

Data mining is important in network planning because it allows the extraction, from data warehouses, data marts, or databases, of information on customers. The information to be mined relates to marketing satisfaction, brand loyalty, predisposition to new services, and other subjective categories. Such information is important for predicting demand for services and facilities. It is also useful for preventive maintenance of facilities, which can range in size from local to global networks. That is, service/equipment failures can be examined to locate faulty hardware/software and diagnose failure patterns. With millions of customers, in many markets, data mining is no trivial task.

Initially, only large companies could afford to do data mining. A typical installation could (and might in some cases still) cost a million dollars. But as of April 1997 more than 50 data mining products have been released, aimed at a variety of users [Edel]. IBM, Thinking Machines, Silicon Graphics, and a number of other vendors are involved in data mining. Prices in 1997 for software ranged from under $1000 for limited desktop systems up to $135,000 and higher for systems running on a variety of platforms [Edel].

Naturally, software alone does not make for a successful data mining project. There are costs associated with the database itself, including ensuring data integrity. There are also costs associated with internal or external (i.e., consulting) expertise needed to implement a data mining system and to provide interpretation/analysis of the results [Nadi].

Before moving on to examine data mining in more detail, it is necessary to point out that the size of a database is not necessarily an indicator of project success. Very successful data mining can be done on databases of small or moderate size [Smal].

6.3.2 Types of Mining

In the usual database query and reporting tools, requests for information need to be so precise that one often must know what one is looking for before beginning the search. Data mining tools, but way of contrast, are more general. Using techniques [Nadi] such as statistics, decision theory, and neural networks, these tools seek to find trends and patterns in the data store that corporate personnel would be likely to request if they know of their existence. A data mining search still needs to be specified, but data mining technology possesses much more flexibility over traditional database query systems to find unexpected trends and patterns.

The pattern types that data mining software can find vary from tool to tool. In discussing the type of pattern that can be found, we will mostly follow the schema of Chen, Han, and Yu [Chen]:

Rule discovery "Rules" [Pars] are relationships between different attributes. The data miner may specify the attributes of interest, and data mining software can find relations between the attributes. Among the types of relationship that can be found are **association rules**, which identify associations between different items. For instance, in each of the following cases of facts brought to light by means of data mining at a large company, unless one knew exactly what one was looking for, the relationship might never have come to light in the course of a search made by means of traditional techniques.

- Modems purchased from a particular vendor during the last quarter of 1996 are three times as likely as other modems to be involved in frequent loss of carrier.

- Customers who usually pay their bills early are significantly more likely than the average customer to subscribe to high end services provided by the company.

- Families with a wide geographic dispersion are more likely to subscribe to access e-mail services.

Sequence discovery This is akin to association analysis except that associations (or patterns) occur over time [Edel].

Generalization and summarization The details of "low concept" data may need to be aggregated into "high concept" data [Chen]. Low concept data, for instance, may be the date, times, and geographic end points of individual long-distance calls. High concept data might be the amount of calls carried on a long-distance fiber optic link by hour, averaged over 30 days. A number of techniques to accomplish generalization and summarization are available [Chen].

Classification and clustering Classificaiton software finds relationships with predictive power in already classified data. Cluster analysis (see Section 6.4.5), on the other hand, partitions data into "clusters" based on their similarity. Unlike classification models, there is no a priori knowledge concerning the nature of the clusters in cluster analysis. Both classification and cluster analysis originated in statistics but have received attention from researchers and developers in artificial intelligence.

Path traversal patterns This is the study, still in its infancy, of usage patterns in hyperlinked media, such as the World Wide Web.

Data anomalies Unusual or rare data patterns could be due to incorrectly entered data or, more interestingly, they could be indicators of interesting phenomena. As Parsaye and Chignell point out [Pars], a large enough number of anomalies may signify the need for rule modification or new rule creation.

Data integrity Guaranteeing data integrity can account for 60% of the up-front work of data mining [Nadi]. Moreover, data preparation is the most time-intensive step in data mining [Edel].

Errors of various sorts can occur in almost any database. Manually entered data in particular is susceptible to occasional errors. Thus there is a definite need for automated error detection because of the large amounts of data usually involved. Such automated processes may enforce constraints, check data dependencies, flag missing data, maintain consistency, and take other steps [Pars]. There is also a need for periodic data quality audits. An organization should have in place processes and policies that promote overall data integrity.

Statistically based techniques used in data mining also include regression, time series analysis, discriminant analysis, and logistic regression [Edel] [Smal] (see Section 6.4). Algorithmically based techniques used in data mining include decision trees and neural networks, as well as data visualization techniques.

Often a number of these various techniques are used on the same problem in a coordinated fashion.

6.3.3 Things To Look for in a Product

The choice of a data mining tool should entail the determination of the availability of a number of features. In categorizing important features, we largely follow Chen, Hand, and Yu [Chen]. It should be noted, however, that these are inherent trade-offs associated with the selection of one or another of the following product features.

Processing different data types Different types of data (alphanumeric, spatial, multimedia, legacy, etc.) are stored, often separately, in different databases. Specific data mining tools may in fact be needed, at this time, for each type.

Data access Some tools require that data be downloaded into their file structure. Others can work directly with a database [Edel].

Algorithmic performance and scalability Many of the algorithms one would like to use in data mining can be extremely computationally expensive if improperly implemented. Thus it is important to utilize efficient implementations. Even an $\mathcal{O}(N^2)$ algorithm[1] may not "scale" well to very large size problems. Because of the need for high performance, parallel processors and parallel memory access can be used in data mining.

Expressions of results High level languages and graphical user interfaces are preferred so that users who are not data mining experts can specify searches and understand results. Uncertainty in results should be clearly (perhaps statistically) indicated.

Interactive mining As anyone who has used a library database or done an Internet search will know, the ability to interactively modify the parameters of a search based on initial results is a very powerful tool. The same ability is very useful in data mining. In fact, data mining is much more useful as an interactive technology than as magic software that will run and immediately produce useful results [Mann 96].

Data fusion Much data is stored in distributed systems interconnected by corporate intranets or even by the Internet. Data mining in such distributed environments, as opposed to data mining in a homogeneous database, is a new challenge.

Related-products interfaces This includes interfaces between data mining tools and databases, data visualization software, and other corporate information systems.

Data security and privacy Corporate data is valuable, and reasonable steps should be taken to preclude theft and tampering. Moreover the ability to do data mining and act on the results must be implemented in a way that does not violate individual privacy and societal mores.

---

[1] Meaning computation and/or memory requirements are proportional to the square of the size of the problem.

6.3.4 Conclusion

Data mining generally leads to incremental but cost-efficient improvements in business operations. The "surprise" that revolutionizes a business can happen, but this is a rarer event. The techniques used in data mining are "natural extensions" [Smal] of data analysis techniques that in some cases date back to the beginnings of the twentieth century. What has dramatically changed is our ability to store large amounts of data and process it in computers.

The reader interested in learning more on data mining is referred to concise, helpful treatments in the literature [Aubr] [Conn] [Edel] [Nadi] [Smal]. A survey with an emphasis on algorithms appears in [Chen]. A good introduction appears in [Mann 96] along with some database issues. Book-length treatments appear in [Piat] [Fayy]. Periodicals such as *Information Week* carry accessible articles on data mining. More in-depth technical articles appear in the *IEEE Transactions on Knowledge and Data Engineering* and the *ACM Transactions on Databases*. The Thinking Machines Corp. Web site at *www.think.com* is particularly helpful as a source of data mining information.

6.4   STATISTICS

6.4.1 Introduction

One cannot talk about data analysis without talking about statistics. In fact, most of the data mining products discussed in Section 6.3 rely heavily on statistical techniques. This section provides a management-level view of important statistical techniques, particularly some of those used in data mining.

Statistics owes much to its older cousin, probability theory, for providing its analytic machinery. Statistics itself can be defined as the analysis of real-world data sets. The study of probability theory began in the mid-1600s with the work of Fermat and Pascal. Its original application was in gambling problems. The science of statistics developed in the 1800s and 1900s. Its original application area was a product of the need to make sense of astronomical data. Later, statistics would continue to develop in response to requirements to analyze data arising in psychology, agriculture, government, industry, science, and technology. In fact the name "statistics" derives from its use in analyzing governmental data (i.e., *state* data). The reader interested in a concise history of statistics is referred to Pansaye and Chignell [Pars].

6.4.2 Regression

The basic problem in **regression** can be intuitively be thought of as one of curve fitting. That is, in the simplest case, one is given a set of m measurements (also called dependent variables) of some quantity, Y_i, $i = 1, 2, \ldots, m$. Each of the Y_i is associated with a quantity called the independent variable, X_i. The independent variable is often time. For instance, Y_i may be the hourly number of calls processed by a switch and X_i may be an identification number for the hour. Data may be taken over a month of switch operation. The typical regression problem is to find a curve and its parameters that well summarize such data. Thus we have a problem in "curve fitting."

A curve to be fit to data can be linear, such as

$$y_i = \sum_{j=1}^{n} \theta_j f_j(t_i) \qquad i = 1, 2, \ldots, m \tag{6.1}$$

where y_i is a measurement at the ith time instant (t_i) and f_j is a ("basis") function of time. Here also, the n parameters, θ_j, $j = 1, 2, \ldots, n$, appear linearly and are to be found for a given data set. The basis functions may be nonlinear functions of time, but the overall model is still considered to be a linear one, since the unknown parameters, θ_j, appear linearly. A nonlinear regression model, on the other hand, may have the form:

$$y_i = \Phi(\theta, t_i) \qquad i = 1, 2, \ldots, m \tag{6.2}$$

Once again n parameters are represented as a parameter vector, θ. The measurements, y_i, depend nonlinearly on the parameters, as well as possible depending nonlinearly on time, t_i.

What is the purpose of fitting a curve to data? One reason is to summarize, simply, the behavior of a potentially large number of data points. A second reason, which is probably more important, is prediction. That is, once we have a curve that fits the data well, we can predict y_i for a given t_i. In the switch example given earlier, for instance, one may wish to predict traffic next year based on this year's data, or perhaps based on data for several years.

Various measures of the "fit" of a curve to data are available. In the common ones listed below, $\Phi(\theta, t_i)$ is used to represent either a linear or nonlinear function.[2]

Least squares error fit

$$\min_{\theta} \left\{ \sum_{i=1}^{m} [y_i - \Phi(\theta, t_i)]^2 \right\}$$

Absolute error fit

$$\min_{\theta} \left[\sum_{i=1}^{m} |y_i - \Phi(\theta, t_i)| \right]$$

---

[2] Linear models are a special (very tractable) case of nonlinear models.

Min-max error fit

$$\min_{\theta} \left[\max_{i=1,2,\ldots,m} |y_i - \Phi(\theta, t_i)| \right]$$

The differences between measurements and model values are called **residuals**.

Particular criteria of fit are optimal in the presence of certain types of random error (e.g., noise). For instance, the use of the least squares estimate is optimal for normally distributed errors. The presence of **outliers**, rogue measurements at extreme values, affects the level of goodness of fit achieved.

Regression may be much more involved than is indicated by the two-dimensional Y/X case. For instance, there may be multiple measurements for each independent unit studied. This is referred to as **multivariate analysis**. For example, each business filling out a questionnaire on telecommunication needs may answer 25 questions (creating 25 dependent variables). When there are many independent variables in the data set, one may be interested in finding the subset of independent variables that best predict the data. Such **variable selection** is useful in initial exploratory data analysis [Afif]. Neural networks can be used to create and solve regression models [Edel].

6.4.3 Discriminant Analysis

The goal in **discriminant analysis** is to identify a particular case as belonging to one of several populations. For example, consider the following two problems.

- Based on past records, a company has profiles of customers who (a) probably will not purchase new services, (b) may purchase new services, and (c) probably will purchase new services. What should be the criteria for assigning a new customer to one of these groups?

- A cellular telephone company with experience in international markets has a profile of two types of potential franchise. One is generally profitable and one is not. An opportunity to invest in a new franchise occurs. How does the company determine which profile this franchise belongs to?

As one might have guessed, some experience in the form of a "training sample" [Afif] is needed to find the criteria by which to tell which population a new case belongs to. Generally discriminant analysis can be used for characterization (ascertaining which population a new case belongs to) or prediction (identifying the attributes that characterize a particular population).

Although multiple populations can be subjected to discriminant analysis, the two-population situation is the most common. In a two-population situation, each population could be distributed normally about differerent means. The **discri-**

mant function in such a case is as simple as a threshold. That is, if a single attribute of a new case is above a threshold value, T, it is classified as belonging to the population with the larger mean. If the attribute is at or below a threshold, the case is classified as belonging to the population with the smaller mean. If there are two populations with two attribute values, then the discriminant function may be a line (if a case is on one side of the line, choose one population; if the case is on the other side of the line, choose the second population). This can be extended to higher dimensions. Note that in discriminant analysis, the attribute values must be continuous.

The theory of discriminant analysis is geared toward choosing such optimal discriminant functions. Among the advantages of discriminant analysis are simple solutions (such as thresholds) and good sensitivity [Smal]. As with other statistical techniques discussed in this chapter, one must be careful about the suitability of distribution assumptions, the presence of outliers, the independence of cases, and the correctness of the population identification of cases in the training sample [Afif].

6.4.4 Logistic Regression

Logistic regression is used to identify a case as belonging to one of *two* populations (as opposed to several populations potentially found in discriminant analysis). Unlike discriminant analysis, logistic regression can be used with both continous-valued and integer-valued attributes. Moreover the attribute variables can be **categorical variables**. That is, an integer may be assigned to indicate each of a number of clear-cut categories. For instance, for the example of the international cellular company, an integer may indicate that the franchise is in a region with (0) low taxes, (1) modest taxes, and (2) high taxes. Logistic regression is frequently used in the medical field.

Largely following the notation of Afifi and Clark [Afif], we start with the posterior probability of a case coming from a particular population of interest. This probability is called the **logistic function**:

$$P_Z = \frac{1}{1 + e^{C-Z}} \tag{6.3}$$

where C is a constant such that $P_Z = 0.5$ when $C = Z$ and Z is a linear discriminant function:

$$Z = \beta_0 + \beta_1 X_1 + \beta_2 X_2 + \beta_3 X_3 + \cdots + \beta_p X_p \tag{6.4}$$

Here the β are constants and the X_i are attribute values (either integer or continuous).

Logistic regression assumes that the natural log of the "odds" (as in "odds of 2 to 1") can be expressed linearly, as

$$\ln_{(\text{odds})} = \alpha + \beta_1 X_1 + \beta_2 X_2 + \beta_3 X_3 + \cdots + \beta_p X_p \tag{6.5}$$

where

$$\text{odds} = \frac{P_Z}{1 - P_Z} \tag{6.6}$$

Thus for instance, if the two populations are "underdeveloped economies" and "developed economies" and the odds are 2:1 that an economy is under-developed, the underdeveloped economies occur with probability of $2/(2+1)$ and odds $2/3/(1 - 2/3)$ or 2:1.

The idea for a linear model for the $\ln_{(\text{odds})}$ comes from the fact that:

$$\ln_{(\text{odds})} = \ln\left(\frac{P_Z}{1 - P_Z}\right) = \ln\left(\frac{\dfrac{1}{1 + e^{C-Z}}}{1 - \dfrac{1}{1 + e^{C-Z}}}\right) \tag{6.7}$$

$$\ln_{(\text{odds})} = \ln\left(\frac{1}{e^{C-Z}}\right) = Z - C \tag{6.8}$$

Equation (6.5) requires no statistical assumption regarding the distribution of the X_i. In logistic regression, we use a training sample on which regression is performed to calculate the coefficients $\alpha, \beta_1, \beta_2, \ldots, \beta_p$.

Once found, the logistic regression model can be used for either character-ization or prediction.

One useful fact results from equation (6.6) and some algebra:

$$P_Z = \frac{\text{odds}}{1 + \text{odds}} \tag{6.9}$$

So

$$P_Z = \frac{\exp(\alpha + \beta_1 X_1 + \beta_2 X_2 + \beta_3 X_3 + \cdots + \beta_p X_p)}{1 + \exp(\alpha + \beta_1 X_1 + \beta_2 X_2 + \beta_3 X_3 + \cdots + \beta_p X_p)} \tag{6.10}$$

and

$$P_Z = \frac{1}{1 + \exp -(\alpha + \beta_1 X_1 + \beta_2 X_2 + \beta_2 X_2 + \beta_3 X_3 + \cdots + \beta_p X_p)} \tag{6.11}$$

This useful results gives the probability that a case belongs to one of two populations (in this case population I). It is known as the **logistic regression equation**.

A major advantage of logistic regression is its flexibility in accommodating different data types, including categorical variables. Another advantage is the relative simplicity of the model. If the underlying population distributions are multivariate normal, then logistic regression is less efficient than discriminant analysis in terms of the number of samples needed to achieve a given precision. On PCs, at least, using the minimal number of variables is important because the computation can be time-consuming [Afif] [Smal].

A particularly readable treatment of discriminant analysis and logistic regression can be found in Afifi and Clark [Afif].

6.4.5 Cluster Analysis

In discriminant analysis and logistic regression one knows for the training sample how members of the training population are classified. **Cluster analysis**, on the other hand, starts with less structure. The basic clustering problem is to take a set of cases and group them into clusters such that cases within a cluster are "similar" and cases in different clusters are "dissimilar."

Typically, each case may consist of a number of different attributes. These are usually represented numerically so that a quantitative distance measure may be used to determine the degree of similarity of any two cases. Typical measures of similarity include Euclidean (squared) distance, absolute distance, the Manhattan metric, and p power distance [Dura].

Algorithms for clustering may start either with a single group of cases that are broken down into clusters or with individual cases that are aggregated into clusters. Algorithms for cluster analysis include enumeration for small problems, heuristics, and dynamic programming, as well as neural networks. The use of nonenumerative techniques is important, since cluster analysis tends to be computation and memory intensive. Clever algorithmic implementations reduce the number of calculations and amount of memory required in several ways [Dura] [Zupa].

Since it is not known a priori what types of cluster may result from an algorithmic analysis, it is important to have a domain expert, look over the results of a cluster analysis.

6.4.6 Time Series

The ability to predict (forecast) quantities such as future traffic demand, outages and breakdowns, and customer preference can be quite valuable. Prediction using statistics is a field in itself. Such techniques as those discussed in this chapter (e.g., curve fitting, least squares estimation, and Kalman filtering) can be used. **Time series models** are popular in their own right, for forecasting applications.

The basic idea in the simplest and most commonly used time series models is that there exists a series of observations (input) at equispaced time instants, $t_k, k = 1, 2, \ldots$. Call these observations $x_k, k = 1, 2, \ldots$. The goal is to predict the observation at time t_k. Our prediction at this time will be called y_k. Then the popular **autoregressive moving average** or **ARMA** model is written as

$$y_k = \sum_{i=1}^{w_\alpha} \alpha_i x_{k-i} + \sum_{j=1}^{w_\beta} \beta_j y_{k-j} \tag{6.12}$$

where

$$\alpha_1 + \alpha_2 + \alpha_3 + \cdots + \alpha_{w_\alpha} = 1.0 \tag{6.13}$$

$$\beta_1 + \beta_2 = \beta_3 + \cdots + \beta_{w_\beta} = 1.0 \tag{6.14}$$

Note that now the forecast is linear weighted combination of w_α past inputs and w_β past forecasts. The number of past inputs and the number of past forecasts are referred to as "window sizes."

If one accepts the validity of such a linear model, the problem becomes one of choosing the coefficients $\alpha_i, i = 1, 2, \ldots, w_\alpha$ and $\beta_j, j = 1, 2, \ldots, w_\beta$ to minimize the prediction error, $y_k - x_k$ in some sense. This can be done in a variety of ways, depending on the error measure used. For instance, a Kalman filter (see Section 6.6) can be used to optimally estimate the α_i and β_j optimally in a least squares sense.

The reader with an electrical engineering background may recognize that ARMA model is quite similar to the infinite impulse response (IIR) filter model of digital signal processing. There x_k is the filter input at time k and y_k is the filter output. Though the linear models appear similar, the application emphases are different for digital signal processing and time series forecasting. In digital signal processing the main goal is to accentuate certain frequency components of a signal and/or to diminish the strength of other components. In forecasting the main goal is to minimize the prediction error. In this way forecasting is really more similar to Kalman filtering than to digital signal processing. However both time series analysis and digital signal processing have a good deal in common in terms of mathematical tools.

Quite a few forecasting models are special cases of the ARMA model. For instance, the **moving average** model is

$$y_k = \sum_{i=1}^{w_\alpha} \alpha_i x_{k-i} \tag{6.15}$$

Here the current forecast is a weighted linear combination of only the preceding w_α observations. Note that, again, the values of α sum to 1. This model is similar to a finite impulse response (FIR) filter model from digital signal processing. A noteworthy special case, a simple average, occurs when $\alpha_i = 1/w_\alpha$. An objection to using a simple average is that old and new observations are weighted equally.

One can naturally implement a fading memory moving average that weighs more recent measurements more heavily. A popular one is **exponential smoothing**:

$$y_k = \varepsilon x_k + (1 - \varepsilon)y_{k-1}, \qquad 0 \le \varepsilon \le 1 \qquad (6.16)$$

That is, the forecast at time t_k is a weighted sum of the preceding forecast and the current observation. Naturally the choice of ε influences the behavior of this predictor.

When past experience has little influence on future behavior, as in a random walk, last value forecasting [Hill 90] can be used:

$$y_k = x_k \qquad (6.17)$$

Finally, we note that in time series analysis, questions arise as to whether a set of observations consists of linear trends, periodic behavior, irregular behavior, or some combination of these. Questions also arise as to whether the underlying time series model is stochastic or deterministic.

Tutorial introductions to forecasting are available [Goul 91] [Hill 90]. An in-depth look at time series analysis, from an econometric viewpoint, is given by Enders [Ende].

6.4.7 Simulation

The student learning methods of statistical (i.e., mathematical) analysis at first may be impressed and overwhelmed by the sheer size of the field and its many subdisciplines. Eventually though, upon gaining proficiency, he or she comes to realize that statistical analysis, though powerful, has its limitations: it is only as good as the validity of its assumptions; there are real limits to the size and types of problems that can be analyzed; and real-world data is often generated by processes that are not mathematically tractable for description and analysis.

Thus as in the sciences, where theory has a long history, there is a need for experimentation when one is dealing with statistical models. Nothing can replace physical experimentation as the ultimate in realism, and its ability to surprise remains unsurpassed. However a middle approach between physical experimentation and statistical analysis that is frequently used is **simulation**.

The computers used in the simulation of actual systems, such as computer and telecommunication networks, are programmed to mimic (simulate) the behavior of a real system such as LAN, a telephone switch or a corporate network. This can be done, since such systems operate under well-understood rules (protocols). The computer program "places" calls, "transmits" packets, and/or "makes" connections according to these rules. Part of such a program will collect data on the simulated system's behavior for statistical processing and display.

Since a simulation is performed in software, it is easy to examine the effects of changing model assumptions, system modifications, and architectural choices.

It is in fact straightforward to construct sophisticated and even baroque models. Also the person performing the simulation does not need analytical skills of the high level required for statistical analysis.

On the downside, simulation of complex systems can be time and memory intensive. Analytical solutions, once found, are often faster to compute. Moreover, for simulation to be useful, an intelligent understanding of its results remain necessary. In fact, if there are many dependent and/or independent quantities, it may be difficult to see transcending patterns in simulation results.

The usual type of simulation of a network is called a **discrete event simulation**. The "discrete events" are occurrences such as packets being transmitted, a buffer receiving a packet, or a call being switched. Simulations can be run to examine **transient effects**. This is behavior that occurs over a short period of time as a result of some event. For example, the behavior of a system when it is first activated usually differs substantially from the behavior of networks over long time periods. Such **steady state behavior** can also be examined.

An important point regarding discrete event simulations is that often they are stochastic in nature. This is because basic inputs (such as packet arrivals and call placements) are generated randomly throughout the use of **pseudorandom number generators**.

The most common pseudorandom number generators are software products that provide randomlike numbers between 0 and 1 according to a uniform distribution. That is, any number between 0 and 1 is equally likely to occur. These randomlike numbers are reproducible each time the program is run if the same "seed" number is used as an input to the pseudorandom number routine. Use a different seed and a different, though statistically similar, sequence of randomlike events will occur in the simulation.

Suppose, for instance, that the probability of a packet arriving on a particular input line in a time slot is 0.7. Then for each time slot a uniform number [between 0 and 1, or $U(0, 1)$ for short] will be generated. For a given slot, if this number is less than 0.7 in value, a packet *will* be made to arrive in the slot. If the number is greater than 0.7 in value, a packet *will not* arrive in the slot.

Random numbers following continuous distributions, other than the uniform distribution, can be generated using a uniform pseudorandom numer generator in the following manner. Let $f(x)$ be the probability density to be simulated and let $F(x)$ be the associated cumulative distribution function, that is,

$$F(x) = \int_{-\infty}^{x} f(z)dz \qquad (6.18)$$

where z is a dummy variable of integration. Then, to generate the random variable x, distributed according to $f(x)$, generate a uniform pseudorandom number y and let

$$x = F^{-1}(y) \qquad (6.19)$$

where F^{-1} is the functional inverse of the cumulative distribution function.

Since pseudorandom quantities are often used in discrete event simulations, the data collected from such a simulation has a random component. With the use of different seeds, the data for simulation runs will be (hopefully slightly) different. Therefore multiple runs can be performed (replication) and the results averaged.

How many runs is enough? Put another way, how does one know the extent of the purely random component in a number of averaged simulation runs? These questions usually are answered through the use of **confidence intervals**. Consider a quantity with a random component, say switch throughput. A 95% confidence interval for switch throughput, for instance, consists of an upper and a lower limit between which there is a 95% probability that switch throughput will fall. Put another way, 95% of the replicated runs, on average, will have switch throughput values falling in the interval defined by the upper and lower limits.

One can define confidence intervals 85, 90, 95, 98, 99 percent and so on. A little thought will show that the larger the percentage, the wider the confidence interval will be. On graphs, confidence intervals are often plotted as so-called error bars above and below each average value, indicating the upper and lower confidence interval limits. Generally, for a given percentage confidence interval, the closer the bars are to the average values, the less random variability is present in the results. Relatively simple formulas and tables for calculating confidence intervals, based on certain assumptions on the random variability, can be found in elementary books on statistics and data analysis.

A **sensitivity analysis** of a simulation can be performed to determine the extent to which performance measures depend on system parameters. Basically, a simulation is run several times, each time using slightly perturbed parameters of interest. In certain cases techniques are available to speed up this analysis [Kuro].

Parallel processors can be used to speed a simulation by replicating runs on different processors. Performing a single run at a time across a parallel processor is more tricky for sometimes communication overhead and serializability limit the speedup that can be obtained.

Simulation is popular because it is a low cost, flexible means of gathering empirical results. The reader interested in reading more on simulation is referred to the literature [Jain] [Kuro].

6.4.8 BEYOND STATISTICS: RULE GENERATION

An important part of data mining is rule generation (discovery). That is, data mining software will generate "rules" that can be used to characterize and/or do predictions on data in a database. These are similar to the rules used in expert systems except that in expert systems "rules" are a software input, whereas in data mining they are an output. To be more formal, such rule generation software will attempt to determine which subset and ranges of independent variables best predicts a dependent variable(s).

Rule generation software usually requests the user to specify the variables of interest and possibly a desired confidence level [Pars]. A 95% confidence level here means that the generated rule or rules account for 95% of the data processed. Naturally, the higher the confidence level percentage, the more elaborate the rules may become.

The careful reader may have noticed that the foregoing statistical techniques could be used either singularly, or in some combination, in rule generation software. Moreover, decision trees [Goul 91] can also be part of this software. **Decision trees** are a logical data structure that allows a hierarchy of decisions to be represented. One of the earliest applications of decision trees was in chess programs.

6.5 LEAST SQUARES MODELING

6.5.1 Introduction

In the field of statistical estimation no theory has been as successfully adopted as that of **least squares estimation (LSE)**. It has been found to well model numerous practical situations. Its mathematics is computationally reasonable. It lends itself to intuition.

What makes the least squares method so successful? One reason is that the underlying model is linear, resulting in a simple, closed-form solution for the estimates. Recursive solutions (i.e., Kalman filters; see Section 6.6) are also possible. A further reason is that the least squares estimate is the maximum likelihood estimator in the presence of additive, zero mean, Gaussian noise. The central limit theorem indicates that many systems in fact can be modeled in this way.

The subsections that follow comprise a tutorial on least squares estimation. The basic model is outlined in Section 6.5.2, followed by a thorough discussion of the structure of the covariance matrix in Section 6.5.3.

6.5.2 The LSE of a Static Model

The concept of choosing parameter estimates that minimize a sum of square errors (i.e., LSE) was independently developed by Gauss in 1795 and by Legendre in 1805 for certain astronomical calculations [Pars] [Sore]. The basic linear measurement model is

$$Y_i = \sum_{j=1}^{n} \theta_j X_{ij} + V_i \qquad i = 1, 2, \ldots, m \qquad (6.20)$$

In matrix form this is

$$Y = X\theta + V \qquad (6.21)$$

$$Y = \begin{bmatrix} Y_1 \\ Y_2 \\ \vdots \\ Y_m \end{bmatrix} \qquad X = \begin{bmatrix} X_{11} & X_{12} & \cdots & X_{1n} \\ X_{21} & X_{22} & \cdots & X_{2n} \\ \vdots & \vdots & \ddots & \vdots \\ X_{m1} & X_{m2} & \cdots & X_{mn} \end{bmatrix} \qquad \theta = \begin{bmatrix} \theta_1 \\ \theta_2 \\ \vdots \\ \theta_n \end{bmatrix} \qquad (6.22)$$

The vector Y consists of m independent measurements. The observed pheonomon is modeled by a linear equation consisting of known constants $X_{ij}, i = 1, 2, \ldots, m, j = 1, 2, \ldots, n$ and unknown parameters $\theta_i, i = 1, 2, \ldots, n$. Thus there are n constant parameters that must be estimated. The model observations are obscured by the presence of an additive and unobservable measurement error (noise) vector, V:

$$V = \begin{bmatrix} V_1 \\ V_2 \\ \vdots \\ V_m \end{bmatrix} \qquad (6.23)$$

The "least squares" estimate of θ is determined from a minimization of sum of the squares of the modeling error. Here "modeling error" is the difference between the actual measurement and the estimated model values. One seeks to

$$\min_{\theta} \left[\sum_{i=1}^{m} \left(Y_i - \sum_{j=1}^{n} \theta_j X_{ij} \right)^2 \right] \qquad (6.24)$$

In matrix form this is

$$\min_{\theta} (Y - X\theta)^T (Y - X\theta) \qquad (6.25)$$

Assume that the variance of the measurement error is normalized to unity. By equating this objective function's first derivative with respect to the parameters, θ, to zero one obtains for the mean estimates of the parameters:

$$\hat{\theta} = (X^T X)^{-1} X^T Y \qquad (6.26)$$

This is the well-known least squares estimate of $\hat{\theta}$. The **covariance** estimate of the parameter estimate is

$$P = (X^T X)^{-1} \qquad (6.27)$$

The matrix $(X^T X)^{-1} X^T$ linearly operates on the measurements, Y, to create the solution.

Covariance matrices, in particular, require some intuitive explanation. A covariance matrix is a square array of entries indicating in the iith entry (i.e., ith row, ith column) the variance of the ith parameter estimate. This is essentially a measure of the random variation (spread) of the estimate about its mean value. The ijth entry $(i \neq j)$ is the covariance of parameter estimate i and j. The covariance can be thought of as a measure of the statistical relationship (loosely "correlation") between the ith and jth parameter estimates.

The least squares estimator is optimal under a variety of criteria. One example is that of **maximum likelihood** [Bick]. Assume that the measurement errors are identical and independent random variables with a normal distribution of zero mean and unity variance. The joint probability density function, if $\mathbf{P} = \mathbf{I}$, is

$$\frac{1}{(2\pi)^{m/2}} \exp\left[\frac{-(\mathbf{Y} - \mathbf{X\theta})^{\mathrm{T}}(\mathbf{Y} - \mathbf{X\theta})}{2}\right] \tag{6.28}$$

This "likelihood" function is parametrized by $\mathbf{\theta}$. It is straightforward to show that it is maximized by the LSE of equation (6.26).

The least squares estimate also lends itself to an elegant geometric interpretation. Let the columns of $\mathbf{X}$ span a subspace called the function space (see Figure 6.1). That is, every point in the space can be formed as a linear combination of the columns of $\mathbf{X}$. The measurement vector, $\mathbf{Y}$, originates in this space (the origin is located here) and extends out of it. Associated with the least squares estimate is the resulting "predicted" measurement vector, which lies within the function space. It is a linear combination of the columns of $\mathbf{X}$. This is

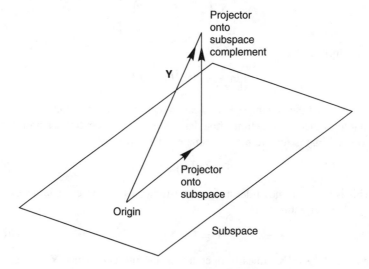

Figure 6.1 Least squares estimation as a projection.

the **projection** of the measurement vector onto the function space. It is well known in estimation theory that the least squares estimate minimizes the magnitude (size) of the projection onto the orthogonal complement of the function space. Intuitively, the predicted measurement vector in the function space is chosen to minimize, in some sense, its distance to the measurement vector, **Y**. The sense of distance is a least squares one.

Put another way, the vector difference between **Y** and the projection onto the function space is the projection onto the function space complement (i.e., space outside the function space). The magnitude of this vector difference is minimized by the least squares estimate. For the least squares estimate, the projection onto the function space and the projection onto the function space complement are at right angles. Along with the measurement vector, a right triangle is formed. These ideas can be stated more formally [Luen 69].

One can use the application of curve fitting to provide an even more intuitive explanation of this. Often it is desirable to fit a curve to a set of noisy real-world data, but no matter how the curve parameters are varied, it is usually not possible to have the curve pass through every data point. This is what is meant by saying that the measurement vector lies "outside" the function space. As one might guess, the function space is the set of all the curves that can be generated by varying the parameters of the curve. If one finds the curve parameters that minimize the sum of the squared errors between the curve and the data points, one has found the least squares estimate of the curve parameters. Thus the error "distance" between the data points and the curve has been minimized. Analogously, the distance between the measurement vector and the function space has been minimized.

For the application of curve fitting one may have a linear **regression function** of the form

$$Y_i = \sum_{j=1}^{n} \theta_j f_j(t_i) + V_i \tag{6.29}$$

The "θ" are the linear parameters to be estimated and the f_i are the $j = 1, 2, \ldots, n$ "basis" functions (or "curves") of time (t_i). Naturally Y_i is the measurement at the ith time sample and V_i is the (unknown) measurement noise at that time. Then, **Y**, **θ**, and **V** are as before, and

$$\mathbf{X} = \begin{bmatrix} f_1(t_1) & f_2(t_1) & \cdots & f_n(t_1) \\ f_1(t_2) & f_2(t_2) & \cdots & f_n(t_2) \\ \vdots & \vdots & \ddots & \vdots \\ f_1(t_m) & f_2(t_m) & \cdots & f_n(t_m) \end{bmatrix} \tag{6.30}$$

where m is, again, the number of measurements. By substituting this **X** into equation (6.26), a least squares fit of this family of curves to the measurement data can be obtained.

There are some other natural extensions of the preceding linear model. Let the variance and covariances of the measurement errors be given by

$$
\mathbf{Q} = \begin{bmatrix}
\sigma_{11}^2 & \sigma_{12}^2 & \cdots & \sigma_{1m}^2 \\
\sigma_{21}^2 & \sigma_{22}^2 & \cdots & \sigma_{2m}^2 \\
\vdots & \vdots & \ddots & \vdots \\
\sigma_{m1}^2 & \sigma_{m2}^2 & \cdots & \sigma_{mm}^2
\end{bmatrix}
\tag{6.31}
$$

The likelihood function then is

$$
\frac{1}{(2\pi)^{m/2}|\mathbf{Q}|^{1/2}} \exp\left[\frac{-(\mathbf{Y} - \mathbf{X}\boldsymbol{\theta})^{\mathrm{T}}\mathbf{Q}^{-1}(\mathbf{Y} - \mathbf{X}\boldsymbol{\theta})}{2}\right]
\tag{6.32}
$$

The "weighted" least squares estimate then is

$$
\hat{\boldsymbol{\theta}} = (\mathbf{X}^{\mathrm{T}}\mathbf{Q}^{-1}\mathbf{X})^{-1}\mathbf{X}^{\mathrm{T}}\mathbf{Q}^{-1}\mathbf{Y}
\tag{6.33}
$$

That is, the estimate of each θ_i is weighted to compensate for the difference, in measurement variances and covariances.

In one frequently occurring specialization, we assume that the measurement errors are independently distributed, with common variance σ^2 and zero correlation. Then

$$
\hat{\boldsymbol{\theta}} = \sigma^2(\mathbf{X}^{\mathrm{T}}\mathbf{X})^{-1}\mathbf{X}^{\mathrm{T}}\mathbf{Y}
\tag{6.34}
$$

Another extension is to allow p sets of m measurements each.

$$
\bar{\mathbf{Y}} = \begin{bmatrix}
Y_{11} & Y_{12} & \cdots & Y_{1p} \\
Y_{21} & Y_{22} & \cdots & Y_{2p} \\
\vdots & \vdots & \ddots & \vdots \\
Y_{m1} & Y_{m2} & \cdots & Y_{mp}
\end{bmatrix}
$$

$$
\bar{\boldsymbol{\theta}} = \begin{bmatrix}
\theta_{11} & \theta_{12} & \cdots & \theta_{1p} \\
\theta_{21} & \theta_{22} & \cdots & \theta_{2p} \\
\vdots & \vdots & \ddots & \vdots \\
\theta_{n1} & \theta_{n2} & \cdots & \theta_{np}
\end{bmatrix}
\tag{6.35}
$$

$$
\bar{\mathbf{V}} = \begin{bmatrix}
V_{11} & V_{12} & \cdots & V_{1p} \\
V_{21} & V_{22} & \cdots & V_{2p} \\
\vdots & \vdots & \ddots & \vdots \\
V_{m1} & V_{m2} & \cdots & V_{mp}
\end{bmatrix}
\tag{6.36}
$$

Now the model equation is

$$\bar{Y} = X\bar{\theta} + \bar{V} \tag{6.37}$$

The least square estimate is then

$$\hat{\bar{\theta}} = (X^T X)^{-1} X^T \bar{Y} \tag{6.38}$$

where

$$Q = \sigma^2 I \tag{6.39}$$

and

$$\sigma^2 = 1.0 \tag{6.40}$$

6.5.3 Structure of the Covariance Matrix

The Covariance Matrix.   If the measurement errors are uncorrelated and the common variance has a normalized value of unity, then the inverse of the covariance matrix is

$$X^T X = \tag{6.41}$$

$$\begin{bmatrix} \sum\limits_{i=1}^{m} X_{i1}^2 & \sum\limits_{i=1}^{m} X_{i1}X_{i2} & \cdots & \sum\limits_{i=1}^{m} X_{i1}X_{in} \\ \sum\limits_{i=1}^{m} X_{i2}X_{i1} & \sum\limits_{i=1}^{m} X_{i2}^2 & \cdots & \cdots \\ \vdots & \vdots & \ddots & \vdots \\ \sum\limits_{i=1}^{m} X_{in}X_{i1} & \cdots & \cdots & \sum\limits_{i=1}^{m} X_{in}^2 \end{bmatrix} \tag{6.42a}$$

This is the normalized **Fisher information matrix,** named after the pioneering statistician. In some applications, such as curve fitting, the $X_{ij}(i = 1, 2, \ldots, m)$ are samples of a jth function at time instants $t_1, t_2, \ldots, t_m$. This would appear as follows:

$$X^T X = \begin{bmatrix} \sum\limits_{i=1}^{m} f_1^2(t_i) & \sum\limits_{i=1}^{m} f_1(t_i)f_2(t_i) & \cdots & \sum\limits_{i=1}^{m} f_1(t_i)f_n(t_i) \\ \sum\limits_{i=1}^{m} f_2(t_i)f_1(t_i) & \sum\limits_{i=1}^{m} f_2^2(t_i) & \cdots & \cdots \\ \vdots & \vdots & \ddots & \vdots \\ \sum\limits_{i=1}^{m} f_n(t_i)f_1(t_i) & \cdots & \cdots & \sum\limits_{i=1}^{m} f_n^2(t_i) \end{bmatrix} \tag{6.42b}$$

The functions may be orthogonal. In this case as the sampling interval, Δ, approaches zero, the off-diagonal elements approach zero and the diagonal elements indicate the relative squared strength of each function.

The X_{ij} $(i = 1, 2, \ldots, m)$ may also be time-shifted versions of a stochastic process with support at discrete instants. The shifts are indexed by j. In this case as $\Delta \to 0$, the diagonal elements represent autocorrelations of overlapping segments of the process. The off-diagonal terms represent cross-correlations of time-shifted versions of the function.

Consider the value of the following two versions of the quadratic form:

$$\mathbf{y^T X^T X y} \tag{6.43}$$

$$\mathbf{z^T z}; \quad \mathbf{z = X y} \tag{6.44}$$

The value of the quadratic form is naturally greater then zero. Thus the information matrix is positive definite. The positive definiteness of the information matrix implies that the matrix eigenvalues[3] are positive.

The matrix inverse of the normalized information matrix is the normalized **covariance matrix**. Its eigenvalues are also positive as

$$\mathbf{(X^T X) y} = \lambda \mathbf{y} \tag{6.45}$$

$$\lambda^{-1} \mathbf{y} = \mathbf{(X^T X)^{-1} y} \tag{6.46}$$

The diagonal elements of

$$\mathbf{(X^T X)^{-1}} \tag{6.47}$$

are the variances of the parameter estimate, that is,

$$\mathbf{(X^T X)}_{ii}^{-1} = \int_{\chi} (\theta_i - \hat{\theta}_i)^2 p(\theta_i) d\theta_i \tag{6.48}$$

where $p(\theta_i)$ is the probability density function of θ_i and χ is the space of possible parameter values. Clearly the square difference between various parameter values and the mean parameter estimate is weighted by the probability of each parameter value occurring and is then integrated (averaged).

The off-diagonal terms are the covariances:

$$\int_{\chi} (\theta_i - \hat{\theta}_i)(\theta_j - \hat{\theta}_j) p(\theta_i, \theta_j) d\theta_i d\theta_j \tag{6.49}$$

where $p(\theta_i, \theta_j)$ is the joint probability density function of θ_i and θ_j. Again, there is an integrated (averaged) weighting of the product of the square differences between the $i(j)$ parameter estimate and possible parameter values.

---

[3] Eigenvalues are characteristic numbers associated with a matrix. See any book on matrix algebra for a discussion of eigenvalues.

Even with this variance/covariance structure, it is often more convenient to utilize aggregate properties of the covariance matrix rather than the set of matrix elements. This is typically done by means of geometric interpretations of the covariance matrix or a scalar-valued function of the covariance matrix. In fact the usual scalar functions also have geometric interpretations.

Consider the joint probability density function of the foregoing linear model:

$$p(\boldsymbol{\theta}) = \frac{1}{(2\pi)^{m/2}|\mathbf{Q}|^{1/2}} \exp\left[\frac{-(\mathbf{Y} - \mathbf{X}\boldsymbol{\theta})^{\mathsf{T}}\mathbf{Q}^{-1}(\mathbf{Y} - \mathbf{X}\boldsymbol{\theta})}{2}\right] \qquad (6.50)$$

A contour of constant probability is given by

$$(\mathbf{Y} - \mathbf{X}\boldsymbol{\theta})^{\mathsf{T}}\mathbf{Q}^{-1}(\mathbf{Y} - \mathbf{X}\boldsymbol{\theta}) = c^2 \qquad (6.51)$$

where c is a constant. This matrix equation describes an ellipsoid-like contour in n-dimensional space (hyperellipsoid). The principal axes of the ellipse can be made coincident with the coordinate axes through diagonalization of the covariance matrix.

Scalar Functions of the Covariance Matrix. It is often desirable to have a single scalar quantity as a measure of the amount of uncertainty represented by a given covariance matrix. That is, given a covariance matrix with its diagonal and off-diagonal terms, a measure summarizing the "size" of the uncertainty is desired. In fact a number of relationships between the covariance matrix and its inverse, their entries, and their eigenvalues are useful for this purpose.

The matrix entries of equation (6.42a) are suggestive of the multidimensional representation of a parallelepiped. In terms of a parallelepiped, the diagonal elements represent the squares of edge lengths, and the off-diagonal elements represent the inner products between edges.

In either case the underlying problem is the selection of a scalar measure that adequately describes the "size" of a parallelepiped (hyperellipsoid). Among commonly used functions (Figure 6.2) are those listed in Table 6.1.

Each function in Table 6.1 has its advantages and disadvantages. The determinant measures the "volume" of the covariance parallelepiped, providing information that can be useful if there are significant off-diagonal terms. On the other hand, the value of the determinant is equally sensitive to changes in each matrix eigenvalue. This can be a problem in that changes in a relatively small eigenvalue have as much effect as changes in a more significant one. However, the determinant has appeared most often in the experimental design literature [Fede] because of its analytical tractability. The trace is useful if off-diagonal terms are not significant. For the trace's additive nature to be meaningful, however, the matrix diagonal elements should be comparable in physical meaning and units.

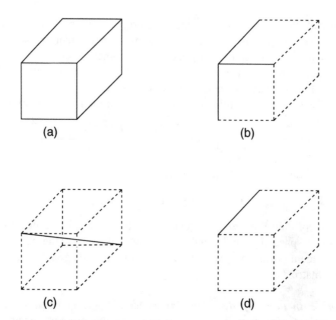

Figure 6.2 How large is a box? (a) Volume or determinant. (b) Sum of edges or trace. (c) Diagonal or frobenius norm. (d) Largest side or maximum eigenvalue.

The last two scalar measures are matrix norms; that is, the matrix norm, $v(\ )$, is such that [Stew]:

1 $A \neq 0 \Rightarrow v(A) > 0$

2 $v(\alpha A) = |\alpha| v(A)$

3 $v(A + B) \leq v(A) + v(B)$

Consistency $v(AB) \leq v(A) v(B)$

TAblE 6.1 COMMONly USEd ScAlAR FUNCTiONS

Measure	Eigenvalues	Matrix Entries[a]
Determinant	$\prod \lambda_i$	Determinant (volume measure)
Trace	$\sum \lambda_i$	$\sum a_{ii}$
Frobenius	$\sqrt{\sum \lambda_i^2}$	$\sqrt{\sum \sum a_{ij}^2}$
Max λ	Max λ	$\|A\|_2 = \max|y^T A x|$ where $\|x\|_2 = \|y\|_2 = 1$ *Note:* $\|A^T A\|_2 = \|A\|_2^2$ *Note:* $\|A^T\|_2 = \|A\|_2$

[a] The λ_i are matrix eigenvalues and the a_{ij} are matrix entries; also $\|\ \|$ is a norm operator (see text).

Intuitively, matrix norms provide a scalar measure of the "size" of a matrix in a systematic and comparable manner.

Some matrix norms, in addition to those above, which may be useful as scalar measures of covariance matrices [Stew], include for an $n \times n$ matrix, **P**, with entries p_{ij}:

$$\|\mathbf{P}\|_{Ab} = \sum_{i=1}^{n} \sum_{j=1}^{n} |p_{ij}| \tag{6.52}$$

$$\|\mathbf{P}\|_1 = \max \left(\sum_{i=1}^{n} |p_{ij}| : j = 1, 2, \ldots, n \right) \tag{6.53}$$

$$\|\mathbf{P}\|_\infty = \max \left(\sum_{j=1}^{n} |p_{ij}| : i = 1, 2, \ldots, n \right) \tag{6.54}$$

$$n \times \max(|P_{ij}| : i = 1, 2, \ldots, n; j = 1, 2, \ldots, n) \tag{6.55}$$

6.6 KALMAN FILTERS

6.6.1 INTRODUCTION

Developed in time to help mankind get to the moon, the Kalman filter is perhaps the greatest twentieth-century success of the field of statistical estimation. What is the Kalman filter? It is a set of equations, based on a certain generic type of statistical model, that allow computers to do filtering, prediction, and smoothing. The equations were developed by R. Kalman in 1958.

"Filtering" here has a particular connotation. The basic idea is that one or more noisy measurement records up to and including time t can be filtered by means of the Kalman equations, to recover an underlying "signal" in the measurements at time t. The design approach used is to develop equations modeling the dynamic evolution of the signal (or "states") and equations modeling the way in which measurements are made.

The reader who has taken a course in signal processing will realize that the design approach used in implementing such a "filter" is somewhat different from the design approach used in frequency-selective filters in signal processing. There the usual design specification is to accentuate and/or diminish certain frequencies. While a Kalman filter may in fact manipulate the frequency spectrum, this is usually an end result (not a specification) of the design process.

"Prediction" involves trying to forecast the signal/state(s) present in noisy measurement data beyond the temporal end of the available data record. "Smoothing" involves estimating the signal/states at time t based on measurements made both before and after time t.

Kalman filters were first used for filtering, estimation, and prediction in the aerospace industry (e.g., for onboard navigation in the Apollo missions). Since then the Kalman filter has been used for data processing in a wide variety of fields, including network planning. For instance, the use of the Kalman filter for telephone network load forecasting was the topic of most of the January 1982 issue of the *Bell System Technical Journal*.

6.6.2 KALMAN FILTERS IN MORE DETAIL

Either discrete time or continuous time models can be used to implement Kalman filters. Let's consider the discrete time version. There is a state (process) matrix equation and a measurement matrix equation in the basic linear discrete time Kalman filter. Let k be the discrete time index. The state equation is [Gelb]

$$\mathbf{x}_k = \mathbf{\Phi}_{k-1}\mathbf{x}_{k-1} + \mathbf{w}_{k-1} \tag{6.56}$$

Expanding this one has

$$\begin{bmatrix} x_1 \\ x_2 \\ \vdots \\ x_n \end{bmatrix}_k = \begin{bmatrix} \Phi_{11} & \Phi_{12} & \cdots & \Phi_{1n} \\ \Phi_{21} & \Phi_{22} & \cdots & \Phi_{2n} \\ \vdots & \vdots & \ddots & \vdots \\ \Phi_{n1} & \Phi_{n2} & \cdots & \Phi_{nn} \end{bmatrix}_{k-1}$$
$$\times \begin{bmatrix} x_1 \\ x_2 \\ \vdots \\ x_n \end{bmatrix}_{k-1} + \begin{bmatrix} w_1 \\ w_2 \\ \vdots \\ w_n \end{bmatrix}_{k-1} \tag{6.57}$$

where there are n "states," $x_1, x_2, \ldots, x_n$. **States** of a system are the basic underlying variables that completely describe the system being modeled. In our linear model, the states evolve from time instant $(k-1)$ to k both deterministically, $(\mathbf{\Phi}_{k-1}\mathbf{x}_{k-1})$ and with an additive process noise component, $\mathbf{w}_{k-1}$. Note that the state transition matrix, $\mathbf{\Phi}_k$, is usually assumed known. The process noise is assumed to be unobservable Gaussian (normal) noise distributed as

$$\mathbf{w}_k \sim N(\mathbf{0}, \mathbf{Q}_k) \tag{6.58}$$

where $\mathbf{Q}_k$ is the covariance matrix of the noise (thus correlated noise samples are allowed).

In a Kalman filter model one may not observe the noisy states directly. Instead, there can be a measurement matrix equation with anywhere from 1 to m measurement (data) streams. One has

$$\mathbf{z}_k = \mathbf{H}_k\mathbf{x}_k + \mathbf{v}_k \tag{6.59}$$

or

$$
\begin{bmatrix} z_1 \\ z_2 \\ \vdots \\ z_m \end{bmatrix}_k = \begin{bmatrix} H_{11} & H_{12} & \cdots & H_{1n} \\ H_{21} & H_{22} & \cdots & H_{2n} \\ \vdots & \vdots & \ddots & \vdots \\ H_{m1} & H_{m2} & \cdots & H_{mn} \end{bmatrix}_k \begin{bmatrix} x_1 \\ x_2 \\ \vdots \\ x_n \end{bmatrix}_k + \begin{bmatrix} v_1 \\ v_2 \\ \vdots \\ v_m \end{bmatrix}_k \tag{6.60}
$$

Here, again, there is a deterministic component ($\mathbf{H}_k\mathbf{x}_k$) and a stochastic component ($\mathbf{v}_k$). The matrix that transforms states to measurements, $\mathbf{H}_k$, is known. The unobservable noise vector, $\mathbf{v}_k$, is distributed according to the normal distribution, $N(\mathbf{0}, \mathbf{R}_k)$. Here $\mathbf{R}_k$ is the measurement noise covariance matrix. Correlated measurement noise is thus allowed. However the process noise, $\mathbf{w}_k$, is assumed to be uncorrelated with the measurement noise, $\mathbf{v}_k$.

Note that the model described by the state and measurement matrix equations is a linear one. Nonlinear filtering is also possible and is discussed in Section 6.6.3.

What Kalman found was that optimal (in the mean square error sense, among others) estimates could be produced by the following set of equations [Gelb]. Assume that measurements are made at discrete time instants, t_k. One has an initial estimate of the state vector, $\hat{\mathbf{x}}_0$ and its covariance, $\mathbf{P}_0$. Then *between* measurements there is a matrix equation for extrapolating the state estimates:

$$
\hat{\mathbf{x}}_k(-) = \boldsymbol{\Phi}_{k-1}\hat{\mathbf{x}}_{k-1}(+) \tag{6.61}
$$

where the "hat" ($^\wedge$) indicates an estimated quantity. A ($-$) indicates a time infinitesimally before a measurement instant and a ($+$) indicates a time infinitesimally after a measurement instant. Thus in equation (6.61) the system goes from the state just after the $(k-1)$st measurement to the state just before the kth measurement. There is also an error covariance extrapolation equation:

$$
\mathbf{P}_k(-) = \boldsymbol{\Phi}_{k-1}\mathbf{P}_{k-1}(+)\boldsymbol{\Phi}_{k-1}^{\mathrm{T}} + \mathbf{Q}_{k-1} \tag{6.62}
$$

These two equations take the system from the time just after the $(k-1)$st measurement to the time just before the kth measurement. Now equations are needed to take the system across a measurement instant. That is, we wish to take the kth time instant measurement vector into account in going from a time just infinitesimally before the kth measurement instant to a time just infinitesimally after the kth measurement instant. Three matrix equations are needed to do this. One is the state estimate update equation:

$$
\hat{\mathbf{x}}_k(+) = \hat{\mathbf{x}}_k(-) + \mathbf{K}_k[\mathbf{z}_k - \mathbf{H}_k\hat{\mathbf{x}}_k(-)] \tag{6.63}
$$

where the $n \times m$ matrix, $\mathbf{K}$, is known as the Kalman "gain" matrix. It is found from:

$$
\mathbf{K}_k = \mathbf{P}_k(-)\mathbf{H}_k^{\mathrm{T}}[\mathbf{H}_k\mathbf{P}_k(-)\mathbf{H}_k^{\mathrm{T}} + \mathbf{R}_k]^{-1} \tag{6.64}
$$

Finally, the last of the necessary three equations is the error covariance update equation:

$$\mathbf{P}_k(+) = [\mathbf{I} - \mathbf{K}_k\mathbf{H}_k]\mathbf{P}_k(-) \tag{6.65}$$

where $\mathbf{I}$ is the identity matrix.

Equation (6.63) is key. We interpret it intuitively to mean that the estimate of the state infinitesimally after the kth measurement instant is equal to the estimate of the state infinitesimally before the kth measurement instant plus a correction term. That correction term is proportional to the size of the difference between the actual kth measurement vector $\mathbf{z}_k$, and the estimated value of the measurement vector, given all the data collected so far, $\mathbf{H}_k\hat{\mathbf{x}}_k(-)$. This error is multiplied by the Kalman gain matrix to produce the correction term.

One of the important advantages of the Kalman filter is its recursive nature. That is, with a reasonable amount of computational effort and memory, it can produce revised estimates of state at every measurement instant. In fact it is *not* necessary to save old measurements, as in least squares "batch" processing. Rather, the estimated state vector and measurement noise covariance matrix are sufficient statistics to summarize the past data.

Kalman filter estimates, if the assumptions are valid, are optimal in several senses, including minimum mean square error, minimum variance unbiased estimation, and maximum likelihood [Gelb].

In **parameter estimation** problems $\mathbf{\Phi}_k = \mathbf{I}$ and $\mathbf{w}_k = \mathbf{0}$ so that $\mathbf{x}_k = \mathbf{x}_{k-1}$. Thus the states (called parameters) are static. The simplified Kalman filter equations in this case are known as the **recursive least squares algorithm**. At the end of processing a finite data record, it produces estimates of the parameters that match that of the least squares batch method for the same data. In parameter estimation applications, the Kalman gain matrix elements, decreases in "size" as the filtering proceeds until convergence and optimal estimation have been achieved.

6.6.3 Implementation Issues

While the form of the Kalman equations presented in Section 6.6.2 is useful analytically, it is not necessarily the best form for implementation on a digital computer. This is because of numerical problems associated with round-off errors. Round-off errors can occur in digital computers making involved calculations because floating point numbers are stored only in a certain finite number of bits. Each addition, subtraction, multiplication, and division may cause some loss in accuracy unless preventive steps are taken. Over a large number of related calculations, these errors can propagate and cause the final estimates to be significantly degraded (i.e., inaccurate). This can happen independent of the correctness of the model and statistical assumptions.

One step that can be taken is to increase the number of bits used to represent numbers. The downside of this is that more memory may be needed (sometimes by a factor of 2 to 4).

A more sophisticated approach to mitigating this problem is to use other mathematical formulations of the Kalman filter equations that are known to have good numerical properties. Kalman filter implementations, based on mathematically sophisticated matrix algebra techniques, including square root filters and UD factorization filters [Brow] [Grew].

For nonlinear models the nonlinear versions of the Kalman filter can be used. These include the extended Kalman filter. The basic idea is to "linearize" the nonlinearity(s) and have a modified filter work in a linearized region. Many real applications of the Kalman filter are actually nonlinear ones.

Filters, be they linear or nonlinear, can fail to converge for a varity of reasons [Grew]. Moreover, a good deal of research effort has been carried out over the years on adaptive filters that can "learn" (adapt) to imperfect knowledge concerning Φ_k, $\mathbf{H}_k$, $\mathbf{R}_k$, and $\mathbf{Q}_k$. Finally, filter equations can be easily modified to include a "fading" memory so that recent measurements are weighted more heavily than older measurements.

6.6.4 Conclusion

Excellent treatments of Kalman filters can be found in the literature [Ande 79] [Brow] [Gelb] [Grew]. In reading these publications, one can seen that using a Kalman filter in an application requires a good knowledge of probability, system theory, and matrix algebra. Yet no other numerical estimation algorithm has such a flexible model, so many applied possibilities and such fast convergence.

6.7 PROBLEMS

1 Why is it beneficial to sometimes display statistics that are within certain ranges?

2 Describe a link map.

3 Describe a 3D global display.

4 Describe a node map.

5 Describe a matrix display.

6 What is data mining?

7 Why is data mining important for network planning?

8 Name and describe two types of pattern that can be found through data mining.

9 Describe the role of errors in data mining.

10 Describe two important features that should be considered in the selection of data mining tool.

11 What is the basic regression problem?

12 Describe the relationship between regression and curve fitting.

13 What measure of fit described in Section 6.4.2 (Regression) do you think is the most generally applicable? Give your reason(s). There is more than one correct answer.

14 Explain multivariate analysis.

15 Describe discriminant analysis.

16 What is the role of a "traning sample" in discriminant analysis?

17 Describe the discriminant function.

18 Name a few pitfalls that may be encountered in doing discriminant analysis.

19 Describe how logistic regression works.

20 Describe the meaning of the multiple logistic regression equation.

21 What is a categorical variable? How is it used in logistic regression?

22 Is discriminant analysis or logistic regression more efficient in terms of the number of samples needed to achieve a given precision if the underlying populations are multivariate normal?

23 What is cluster analysis? Why is it different from discriminant analysis or logistic regression?

24 Describe the role of a quantitative distance measure in cluster analysis.

25 Describe the main ways in which cluster analysis algorithms work.

26 Why is it important to have a domain expert look over the results of a cluster analysis?

27 What is a time series?

28 What is the difference between an ARMA and moving average time series model?

29 Describe the relation between digital signal processing models and time series models.

30 What is exponential smoothing?

31 Describe the role of simulation.

32 What is the difference between a simulation of transient effects and a simulation of steady state behavior?

33 Describe how uniform density pseudorandom number generators are used in simulation.

34 How can (continuous) pseudorandom numbers with a nonuniform density be generated?

35 What is a confidence interval?

36 What is a sensitivity analysis in a simulation?

37 Describe how rule generation works.

38 What is a confidence level in rule generation?

39 What are decision trees (as used in rule generation)?

40 What makes the least squares method so successful?

41 How old is the least squares method?

42 What is the significance of equation (6.24)?

43 What is the role of the covariance matrix in the least squares method?

44 Describe a geometric interpretation of the least squares method.

45 Describe a curve fitting interpretation of the least squares method.

46 What is the significance of equation (6.36)?

47 What is the Fisher information matrix? How is it related to the covariance matrix?

48 Which of the scalar measures of Figure 6.2 do you find the most reasonable? Why? (There is no one correct answer.)

49 What is the problem with the determinant measure in terms of eigenvalue sensitivity?

50 What is a Kalman filter?

51 What is the difference between the measurement model and the process model in a Kalman filter?

52 Name two solutions for mitigating numerical problems caused by round-off errors in Kalman filters.

53 Create a Kalman filter model for estimating the parameters of an ARMA time series model.

BIBLIOGRAPHY

[Adam]   D. Adamo, G. Nocera, and A. Voglino, "RAMSES: A Telecommu-
            nication Network Planning Expert System," In *Proceedings of the
            IASTED International Symposium on Expert Systems: Theory and
            Applications*, Zurich, Switzerland, June 1989. USA Acta Press,
            Anaheim, CA, 1989.

[Afif]   A. A. Afifi and V. Clark, *Computer-Aided Multivariate Analysis*, 3rd
            ed. Chapman & Hall, London, 1996.

[Aho]   A. V. Aho, J. E. Hopcroft, and J. D. Ullman, *The Design and
            Analysis of Computer Algorithms*. Addison-Wesley, Reading, MA,
            1976.

[Ande 79]   B. D. O. Anderson and J. B. Moore, *Optimal Filtering*. Prentice-Hall,
            Englewood Cliffs, NJ, 1979.

[Ande 93]   C. A. Anderson, K. Fraughnaugh, M. Parker, and J. Ryan, "Path
            Assignment for Call Routing: An Application of Tabu Search,"
            Annals of Operations Research, Vol. 41, No. 1–4, 1993, pp. 301–312.

[Appe]   M. Appeldorn, R. Kung, and R. Saracco, "TMN + IN = TINA,"
            IEEE Communications Magazine, March 1993, pp. 78–85.

[Aubr]   D. Aubrey, "Mining for Dollars," *Computer Shopper*, August 1996,
            pp. 568–572.

[Bane]   S. Banerjee and V. O. K. Li, "Order-P: An Algorithm to Order
            Network Partitioning," *Proceedings of IEEE INFOCOM'92*. IEEE
            Computer Society Press, Los Alamitos, CA, 1992, pp. 432–436.

[Barq]   R. Barquin and H. Edelstein, eds., *Planning and Designing the Data
            Warehouse*, Prentice Hall, Upper Saddle River, NJ, 1997.

[Baza]   M. S. Bazaraa, J. J. Jarvis, and H. D. Sherali, *Linear Programming
            and Network Flows*, 2nd ed. Wiley, New York, 1990.

[Beck 91]   R. A. Becker, S. G. Eick, and A. R. Wilks, "Basics of Network
            Visualization," *IEEE Computer Graphics and Applications*, May
            1991, pp. 12–14.

[Beck 95]   R. A. Becker, S. G. Eick, and A. R. Wilks, "Visualizing Network Data," *IEEE Transactions on Visualization and Computer Graphics*, Vol. VCG-1, No. 1, March 1995, pp. 16–28.

[Bell]   Bell Laboratories, "High Capacity Mobile Telephone System Technical Report," Dec 1971, submitted to the Federal Communications Commission.

[Bell 57]   R. E. Bellman, *Dynamic Programming*. Princeton University Press, Princeton NJ, 1957.

[Bell 62]   R. E. Bellman and S. E. Dreyfus, *Applied Dynamic Programming*. Princeton University Press, Princeton, NJ, 1962.

[Bert 81]   J. Bertin, *Graphics and Graphic Information-Processing*. Walter de Gruyter, Berlin, 1981.

[Bert 87]   D. Bertsekas and R. Gallager, *Data Networks*. Prentice-Hall, Englewood Cliffs, NJ, 1987.

[Bick]   P. J. Bickel and K. A. Doksum, *Mathematical Statistics*. Holden-Day, San Francisco, 1977.

[Bogd]   Z. R. Bogdanowicz and T. G. Moore, "A New Optimal Packing Algorithm for Telecommunications Networks Planning," *Computers and Mathematical Applications*, Vol. 18, No. 8, 1989, pp. 739–744.

[Boll]   B. Bollobás, *Random Graphs*. Academic Press, London, 1985.

[Boor]   R. R. Boorstyn and H. Frank, "Large-Scale Network Topological Optimization," *IEEE Transactions on Communications*, Vol. COM-25, No. 1, January 1977.

[Bras]   P. J. Braspenning, F. Thuijsman, and A. J. M. M. Weijters, eds., *Artificial Neural Networks: An Introduction to ANN Theory and Practice*, Lecture Notes in Computer Science no. 931. Springer-Verlag, Berlin, 1995.

[Brow]   R. G. Brown and P. Y. C. Hwang, *Introduction to Random Signals and Applied Kalman Filtering*, 2nd ed. Wiley, New York, 1992.

[Brys]   A. E. Bryson Jr. and Y.-C. Ho, *Applied Optimal Control*. Halstead Press, Wiley, New York, 1975.

[Bstj]   *Bell System Technical Journal*, Special issue on telephone network load forecasting, Vol. 61, No. 1, January 1982.

[Budw]   J. N. Budwey and A. Salameh, "From LANs to GANs," *Telecommunications*, Vol. 26, No. 7, July 1992, pp. 23–27.

[Carl]   S. Carlson, "Algorithm of the Gods," in The Amateur Scientist column, *Scientific American*, March 1997, pp. 121–123.

[Čern]   V. Černy, "Thermodynamical Approach to the Traveling Salesman Problem: An Efficient Simulation Algorithm," *Journal of Optimization Theory and Applications*, Vol. 45, 1985, pp. 41–51.

[Chen]   M.-S. Chen, J. Han, and P. S. Yu, "Data Mining: An Overview from a Database Perspective," *IEEE Transactions on Knowledge and*

Data Engineering, Vol. KDE-8, No. 6, December 1996, pp. 866–883.

[Clel] D. J. Cleland and W. R. King, *System Analysis and Project Management*, 3rd ed. McGraw-Hill, New York, 1993.

[Coad] P. Coad and E. Yourdon, *Object Oriented Design*. Prentice-Hall International, New York, 1991.

[Coch] J. I. Cochrane, W. E. Falconer, V. S. Mummert, and W. E. Strich, "Latest Network Trends," *IEEE Communications Magazine*, Vol. 23, No. 10, October 1985, pp. 17–31.

[Codd] E. F. Codd, "A Relational Model for Large Shared Data Banks," *Communications of the ACM*, Vol. 13, 1970, pp. 377–387.

[Conn] L. Connor, "Mining for Data," *Communications Week*, February 12, 1996, pp. 37–41.

[Coop 74] L. Cooper and D. Steinberg, *Methods and Applications of Linear Programming*. Saunders, Philadelphia, 1974.

[Coop 84] A. Cooper, "Long-Range Planning for Subscriber Network," *Telecommunications Journal*, Vol. 51, July 1984, pp. 363–367.

[Cord] E. S. Cordingley and H. Dai, "Encapsulation—An Issue for Legacy Ssytems," *BT Technology Journal*, Vol. 11, No. 3, July 1993.

[Crab 94] I. B. Crabtree and D. Munaf, "Planning Beneficial and Profitable Network Upgrade Paths," *BT Technology Journal*, October 1994, pp. 92–97.

[Crab 95] I. B. Crabtree, "Resource Scheduling: Comparing Simulated Annealing with Constraint Programming," *BT Technology Journal*, Vol. 13, No. 1, January 1995.

[Crav] H. Cravis, *Communication Network Analysis*, Lexington Books, D.C. Heath, Boston, 1981, pp. 35–50.

[Cusa] E. L. Cusack and E. S. Cordingley, "Object Orientation in Communication Engineering," *BT Technology Journal*, Vol. 11, No. 3, July 1993, pp. 9–17.

[Dant] G. B. Dantzig, *Linear Programming and Extensions*. Princeton University Press, Princeton, NJ, 1963.

[Davi] L. Davis, *Handbook of Genetic Algorithms*. Van Nostrand Reinhold, New York, 1991.

[Dorf] R. C. Dorf, *The Electrical Engineering Handbook*. CRC Press, Boca Raton, FL, 1993.

[Dran] W. Drane and R. H. Barham, "An Algorithm for Least Squares Estimation of Nonlinear Parameters When Some of the Parameters Are Linear," *Technometrics*, Vol. 14, No. 3, 1972, pp. 757–766.

[Dura] B. S. Duran and P. L. Odell, *Cluster Analysis: A Survey*, Lecture Notes in Economics and Mathematical Systems, no. 100. Springer-Verlag, Berlin, 1974.

[Durk]   J. Durkin, "Expert Systems: A View of the field," *IEEE Expert Intelligent Systems and Their Applications*, Vol. 11, No. 2, April 1996, pp. 56–63.

[Edel]   H. Edelstein, "Mining for Gold," *Information Week*, April 21, 1997, pp. 53–70.

[Eick]   S. G. Eick, "Aspects of Network Visualization," *IEEE Computer Graphics and Applications*, March 1996, pp. 69–72.

[Elsa]   E. A. Elsayed, *Reliability Engineering*. Addison-Wesley, Reading, MA., 1996.

[Elia]   D. Elias and M. J. Ferguson, "Topological Design of Multipoint Teleprocessing Networks," *IEEE Transactions on Communications*, Vol. COM-22, November 1974, pp. 1753–1762.

[Ende]   W. Enders, *Applied Econometrics Time Series*. Wiley, New York, 1995.

[Esau]   L. R. Esau and K. C. Williams, "On Teleprocessing System Design, Part II," *IBM System Journal*, Vol. 5, No. 3, 1966, pp. 142–147.

[ESRI]   Environmental Systems Research Institute Inc., *Understanding GIS: The ARC/INFO Method, Self-Study Workbook, PC Version*. Copublished by ESRI Inc., Redlands, CA, and Wiley, Somerset, NJ, 1995.

[Ever]   H. Everett III, "Generalized Lagrange Multipliers Method for Solving Problems of Optimal Allocation of Resources," *Operations Research*, Vol. 2, 1963, pp. 399–418.

[Farr]   C. Farrell, M. Schulze, S. Pleitner, and M. Songerwala, "Network Monitoring and Visualization," *Internetworking: Research and Experience*. Wiley, Chichester, Vol. 6, No. 4, December 1995, pp. 168–183.

[Fayy]   U. M. Fayyad, G. Piatetsky-Shapiro, P. Smyth, and R. Uthurusamy, *Advances in Knowledge Discovery and Data Mining*. AIAA Press/The MIT Press, Cambridge, MA., 1996.

[Fede]   V. V. Federov, *Theory of Optimal Experiments*. Academic Press, New York, 1972.

[Feld]   E. Feldman, F. A. Lehner, and T. L. Ray, "Warehouse Locations Under Continuous Economies of Scale," *Management Science*, Vol. 12, May 1966, pp. 670–684.

[Fish]   M. Fisher, T. Rea, A. Swanton, M. Wilkinson, and D. Wood, "Techniques for Automated Planning of Access Networks," *BT Technology Journal*, Vol. 14, No. 2, April 1996, pp. 121–127.

[Foge]   D. B. Fogel, *Evolutionary Computation: Toward a New Philosophy of Machine Intelligence*. IEEE Press, Piscataway, NJ, 1995.

[Font]   F. J. Fontes, "Planning in the Subscriber Network Maintenance Department," *Telecommunication Journal*, Vol. 52, November 1985, pp. 613–620.

[Ford]   L. R. Ford and D. R. Fulkerson, *Flows in Networks*. Princeton University Press, Princeton, NJ, 1962.

[Foss]   L. D. Fossett, A. Kashper, S. Civanlar, and W. D. Radwill, "Evolution of Routing and Capacity Management in International Switched Networks," *AT&T Technical Journal*, Vol. 71, September/October 1992, pp. 13–23.

[Fox]   B. Fox, "Discrete Optimization Via Marginal Analysis," *Management Science*, Vol. 13, November 1966, pp. 210–216.

[Fran 72]   H. Frank and W. Chou, "Topological Optimization of Computer Networks." *Proceedings of the IEEE*, November 1972, pp. 1385–1397.

[Fran 74]   R. L. Francis and J. A. White, *Facility Layout and Location: An Analytical Approach*. Prentice-Hall, Englewood Cliffs, NJ, 1974.

[Free]   E. Freeman, "Birth of a Terabyte Data Warehouse," *Datamation*, Vol. 43, No. 4, April 1997, pp. 80–84.

[Früh]   T. Frühwirth, P. Brissett, and J.-R. Molwitz, "Planning Cordless Business Communication Systems," *IEEE Expert Intelligent Systems and Their Applications*, Vol. 11, No. 1, February 1996, pp. 50–55.

[Fuss]   J. B. Fussell, "Fault Tree Analysis: Concepts and Techniques," In *Proceedings of the NATO Advanced Study Institute on Generic Techniques in System Reliability Assessments*, E. Henley and J. Lynn, eds., Noordhoff, Leiden, Netherlands, 1976, pp. 133–162.

[Gari]   H. de Garis, "CAM-Brain: The Evolutionary Engineering of a Billion Neuron Artificial Brain by 2001 Which Grows/Evolves at Electronic Speeds Inside a Cellular Automata Machine (CAM)." In *Towards Evolvable Hardware: The Evolutionary Engineering Approach*, E. Sanchez and M. Tomassini, eds. Springer-Verlag, Berlin, 1996, pp. 76–98.

[Gelb]   A. Gelb, ed, *Applied Optimal Estimation*. MIT Press, Cambridge, MA, 1974.

[Gerl 73]   M. Gerla, *The Design of Store and Forward Networks for Computer Communications*. Ph.D thesis, School of Engineering and Applied Science, UCLA, Los Angeles, 1973.

[Gerl 77]   M. Gerla and L. Kleinrock, "On the Topological Design of Distributed Computer Networks," *IEEE Transactions on Communications*, Vol. COM-25, No. 1, January 1977.

[Gilb]   W. E. Gilbert, "The Five Challenges of Managing Global Networks," *IEEE Communications Magazine*, October 1992, pp. 78–82.

[Gles]   M. Glesner and W. Pöchmüller, *Neurocomputers: An Overview of Neural Networks in VLSI*. Chapman & Hall, London, 1994.

[Glov 86]   F. Glover, "Future Paths for Integer Programming and Links to Artificial Intelligence," *Computers and Operations Research*, Vol. 13, No. 5, 1986, pp. 533–549.

[Glov 90]   F. Glover, "Tabu Search: A Tutorial," *Interfaces*, Vol. 4, July/August 1990, pp. 74–94.

[Glov 93a]   F. Glover, M. Laguna, E. Taillard, and D. De Werra, eds., *Tabu Search*, special issues of *Annals of Operations Research*, Vol. 41, No. 1–4. J. C. Baltzer Science Publishers, Basel, Switzerland, 1993.

[Glov 93b]   F. Glover, E. Taillard, and D. De Werra, eds., "A User's Guide to Tabu Search," *Annals of Operations Research*, Vol. 41, No. 1–4. J. C. Baltzer Science Publishers, Basel, Switzerland, 1993, pp. 3–28.

[Goff]   L. Goff, "Network Planners–Architects: Directing a Company's Future," *Computerworld*, Vol. 26, December 21, 1992, p. 61.

[Gold 83]   A. Goldberg and D. Robson, *Smalltalk-80: The Language and Its Implementation*. Addison-Wesley, Reading, MA, 1983.

[Gold 89]   D. E. Goldberg, *New Genetic Algorithms in Search, Optimization and Machine Learning*. Addison-Wesley, Reading, MA, 1989.

[Golu]   G. H. Golub and V. Pereyra, "The Differentiation of Pseudo-Inverses and Nonlinear Least Squares Problems Whose Variables Separate," *SIAM Journal on Numerical Analysis*, Vol. 10, No. 2, 1973, pp. 413–422.

[Goul 87]   E. P. Gould and C. D. Pack, "Communications Network Planning in the Evolving Information Age," *IEEE Communications Magazine*, Vol. 25, No. 9, September 1987, pp. 22–30.

[Goul 91]   F. J. Gould, G. D. Eppen, and C. P. Schmidt, *Introductory Management Science*, 3rd ed. Prentice-Hall, Englewood Cliffs, NJ, 1991.

[Grad]   I. S. Gradshteyn and M. Ryzhik, *Table of Integrals, Series and Products*. Academic Press, New York, 1965.

[Grew]   M. S. Grewal and A. P. Andrews, *Kalman Filtering: Theory and Practice*. Prentice-Hall Englewood Cliffs, NJ, 1993.

[Gros]   D. Gross and C. M. Harris, *Fundamentals of Queueing Theory*. Wiley, New York, 1974, 1985.

[Grov]   W. D. Grover, T. D. Bilodeau, and B. D. Venables, "Near Optimal Spare Capacity Planning in a Mesh Restorable Network," *Proceedings of IEEE Globecom '91*, 1991, pp. 2007–2012.

[Gupt]   V. P. Gupta, "What Is Network Planning?" *IEEE Communications Magazine*, Vol. 23, No. 10, October 1985, pp. 10–16.

[Hand]   G. Y. Handler and P. B. Mirchandani, *Location on Networks: Theory and Algorithms*. MIT Press, Cambridge, MA, 1979.

[Hao]   Q. Hao, B.-H. Soong, E. Gunawan, J.-T. Ong, C.-B. Soh, and Z. Li, "A Low-Cost Cellular Mobile Communication System: A Hierarchical Optimization Network Resource Planning Approach," *IEEE*

Journal on Selected Areas in Communications, Vol. 15, No. 7, September 1997, pp. 1315–1326.

[Hayk]   S. Haykin, *Neural Networks: A Comprehensive Foundation*. IEEE Press, Piscataway, NJ, and Prentice-Hall, Englewood Cliffs, NJ, 1997.

[Hedb]   S. R. Hedberg, "AI's Impact in Telecommunications—Today and Tomorrow," *IEEE Expert Intelligent Systems and Their Applications*, Vol. 11, No. 1, February 1996, pp. 6–9.

[Hill 90]   F. S. Hillier and G. I. Lieberman, *Introduction to Operations Research*, 5th ed. McGraw-Hill, New York, 1990.

[Hill 93]   D. W. Hillis, "Co-evolving Parasites Improve Simulated Evolution as an Optimization Procedure. In *Artificial Life II*, C. Langton et al., eds., Vol. 1, Addison-Wesley, Reading, MA, 1993, pp. 101–125.

[Holl]   J. H. Holland, *Adaptation in Natural and Artificial Systems*. University of Michigan Press, Ann Arbor, 1975.

[Hugh]   J. G. Hughes, *Database Technology: A Software Engineering Approach*. Prentice-Hall, London, 1988.

[Hui]   J. Y. Hui, *Switching and Traffic Theory for Integrated Broadband Networks*. Kluwer Academic Publishers, Norwell, MA, 1990.

[Ishi]   N. Ishiwa and H. Okazaki, "Development of an Enhanced Private Network Planning Expert System: PRINCESS," *NEC Technical Journal*, Vol. 46, No. 7, pp. 79–84 (in Japanese).

[Jack 90]   P. Jackson, *Introduction to Expert Systems*, 2nd ed. Addison-Wesley, Reading, MA, 1990.

[Jack 92]   M. Jackson, *Understanding Expert Systems Using Crystal*. Wiley, Chichester, 1992.

[Jain]   R. Jain, *The Art of Computer System Performance Analysis: Techniques for Experimental Design, Measurement, Simulation and Modeling*. Wiley, New York, 1991.

[Jazw]   A. Jazwinski and B. Laszakovits, "Applying Simulation to Network Planning," *Network World*, Vol. 10, No. 20, May 17, 1993, pp. 33–35.

[Kapu]   K. C. Kapur and L. R. Lamberson, *Reliability in Engineering Design*. Wiley, New York, 1977.

[Karo]   M. Karoński and Z. Palka, *Random Graphs '85*. North-Holland, Amsterdam, 1987.

[Kauf]   A. Kaufmann and A. Henry-Labordère, *Integer and Mixed Programming: Theory and Applications*. Academic Press, New York, 1977.

[Kers]   A. Kershenbaum, *Telecommunication Network Design Algorithms*. McGraw-Hill, New York, 1993.

[Khal]   K. M. Khalil and P. A. Spencer, "A Systematic Approach for

Planning, Tuning and Upgrading Local Area Networks," *Proceedings of IEEE Globecom '91*, 1991, pp. 658–663.

[King]   W. R. King and G. Premkumar, "Key Issues in Telecommunications Planning", *Information and Management*, Vol. 17. Elsevier North-Holland, Amsterdam, 1989, pp. 255–265.

[Kirk]   S. Kirkpatrick, C. D. Gelatt Jr., and M. P. Vecchi, "Optimization by Simulated Annealing," *IBM Research Report RC9355*, 1982. See also the article of the same title and authors in *Science*, Vol. 220, 1983, pp. 671–680.

[Klei 64]   L. Kleinrock, *Stochastic Message Flow and Delay*. McGraw-Hill, New York, 1964.

[Klei 70]   K. Kleinrock, "Analysis and Simulation Methods in Computer Network Design," In *Conference Record of Spring Joint Computer Conference*, AFIPS Conference Proceedings, Vol. 36. AFIPS Press, Montvale, NJ, 1970, pp. 568–579.

[Klei 75]   L. Kleinrock, *Queueing Systems*, Vol. I; *Theory*. Wiley, New York, 1975.

[Kobr]   H. Kobrinski, "Crossconnection of WDM High Speed Channels," *Electronics Letters*, Vol. 23, No. 18, August 1987.

[Kokh]   A. A. Kokhar, V. K. Prasanna, M. E. Shaaban, and C.-L. Wang, "Heterogeneous Computing: Challenges and Opportunities," *(IEEE) Computer*, June 1993, pp. 18–27.

[Kort]   G. B. Korte, *The GIS Book*, 3rd ed, OnWord Press, Sante Fe, NM, 1994.

[Krus]   J. B. Kruskal, "On the Shortest Spanning Subtree of a Graph and the Traveling Salesman Problem," *Proceedings of the American Mathematical Society*, Vol. 7, 1956.

[Kueh]   A. A. Kuehn and M. J. Hamburger, "A Heuristic Program for Locating Warehouses," *Management Science*, Vol. 9, 1963, pp. 643–666.

[Kuma]   H. Kumamoto and E. J. Henley, *Probabilistic Risk Assessment and Management for Engineers and Scientists*, 2nd ed, IEEE Press, Piscataway, NJ, 1996.

[Kuro]   J. F. Kurose and H. T. Mouftah, "Computer-Aided Modeling, Analysis and Design of Communication Networks," *IEEE Journal on Selected Areas in Communications*, Vol. 6, No. 1, January 1988, pp. 130–145.

[Lee 86]   W. C. Y. Lee, "Elements of Cellular Mobile Radio Systems," *IEEE Transactions on Vehicular Technology*, Vol. VT-35, May 1986, pp. 48–56.

[Lee 89a]   M. Lee, "Least-Cost Network Topology Design for a New Service Using Tabu Search," *Heuristics for Combinatorial Optimization*,

Section 6, 1989, pp. 1–18.

[Lee 89b]   W. C. Y. Lee, *Mobile Cellular Telecommunications Systems.* McGraw-Hill, New York, 1989.

[Leem]   L. M. Leemis, *Reliability: Probabilistic Models and Statistical Methods.* Prentice-Hall, Englewood Cliffs, NJ, 1995.

[Leif]   L. J. Leifman, "Estimation of Resource Requirements in Optimization Problems of Network Planning," *European Journal of Operations Research,* Vol. 2, 1978, pp. 265–280.

[Li 84]   V. O. K. Li and J. A. Silvester, "Performance Analysis of Networks with Unreliable Components," *IEEE Transactions on Communications,* Vol. COM-32, No. 10, October 1984, pp. 1105–1110.

[Lieb]   J. Liebowitz, "Expert Systems: A Short Introduction," *Engineering Fracture Mechanics,* Elsevier Science, Amsterdam, Vol. 50, No. 5/6, pp. 601–607, 1995.

[Liew]   S. C. Liew and K. W. Lu, "A Framework for Network Survivability Characterization," *Proceedings of IEEE INFOCOM'92.* IEEE Computer Society Press, Los Alamitos, CA, 1992, pp. 405–410.

[Luen 69]   D. G. Luenberger, *Optimization by Vector Space Methods.* Wiley, New York, 1969.

[Luen 73]   D. G. Luenberger, *Introduction to Linear and Nonlinear Programming.* Addison-Wesley, Reading, MA, 1973.

[Luss]   H. Luss, "A Network Flow Approach for Capacity Expansion Problems with Two Facility Types," *Naval Research Logistics Quarterly,* 1980, pp. 597–608.

[MacD]   V. H. MacDonald, "The Cellular Concept," *Bell System Technical Journal,* Vol. 58, January 1979, pp. 15–42.

[Mann 67]   A. S. Manne, *Investments for Capacity Expansioin: Size, Location and Time Phasing.* MIT Press, Cambridge, MA, 1967.

[Mann 96]   H. Mannila, "Data Mining: Machine Learning, Statistics and Databases." In *Proceedings of the 8th International Conference on Scientific and Statistical Database Management,* P. Svensson and J. C. French, eds., June 1996.

[Marq]   D. W. Marquardt,"An Algorithm for Least Squares Estimation of Non-Linear Parameters," *SIAM Journal,* Vol. II, No. 2, 1963, pp. 431–441.

[Mcca]   J. D. McCabe, *Practical Computer Network Analysis and Design.* Morgan Kaufman Publishers, San Francisco, 1998.

[Mehr]   A. Mehrotra, *Cellular Radio Performance Engineering.* Artech House, Boston, 1994.

[Merc]   A. Merchant and B. Sengupta, "Assignment of Cells to Switches in PCS Networks," *IEEE Transactions on Networking,* Vol NET-3, No. 5, October 1995, pp. 521–526.

[Metr]   N. Metropolis, A. W. Rosenbluth, M. N. Rosenbluth, A. H. Teller, and E. Teller, "Equation of State Calculations by Fast Computing Machines," *Journal of Chemical Physics*, Vol. 21, 1953, pp. 1087–1092.

[Meye]   B. Meyer, *Object Oriented Software Construction*. Prentice-Hall, Englewood Cliffs, NJ, 1988.

[Mich]   Z. Michalewicz, *Genetic Algorithms + Data Structures = Evolution Programs*, 2nd ed. Springer-Verlag, Berlin, 1994.

[Muel]   B. Mueller and J. Reinhardt, *Neural Networks: An Introduction*. Springer-Verlag, New York, 1990.

[Nadi]   L. Nadile, "Data Mining You Can Afford," *Information Week*, March 24, 1997, pp. 88–95.

[Nara]   T. V. Narayana, *Lattice Path Combinatorics with Statistical Applications*. University of Toronto Press, Toronto, 1979.

[Newp]   K. T. Newport and M. A. Schroeder, "Network Survivability Through Connectivity Optimization," *Proceedings of IEEE International Conference on Communications '87*, 1987, pp. 471–477.

[Noum]   P. Noumba Um, "New Prospects in Rural Telecommunications Planning," *Telecommunications Journal*, Vol. 56, September 1989, pp. 582–585.

[Oett]   J. Oetting, "Cellular Mobile Radio—An Emerging Technology," *IEEE Communications Magazine*, Vol. 21, No. 8, November 1983, pp. 10–15.

[Oliv]   S. Oliveira and G. Stroud, "A Parallell Version of Tabu Search and the Path Assignment Problem." *Heuristics for Combinatorial Optimization*, Section 4, 1989, pp. 1–24.

[Osbo]   M. R. Osborne, "Some Special Nonlinear Least Squares Problems," *SIAM Journal of Numerical Analysis*, Vol. 12, No. 4, 1975, pp. 571–592.

[Otte]   R. H. J. M. Otten and L. P. P. P. van Ginneken, *The Annealing Algorithm*. Kluwer Academic Publishers, Boston, 1989.

[Pal 83]   N. E. Pal and M. Fourier, "Trans-Pacific Telecommunications Network Planning," *Telecommunication Journal*, Vol. 50, August 1983, pp. 415–419.

[Pal 85]   N. E. Pal, "An Approach to the Planning of International Telecommunication Networks," *Telecommunications Journal*, Vol. 52, January 1985, pp. 32–42.

[Pars]   K. Parsaye and M. Chignell, *Intelligent Database Tools and Applications*. Wiley, New York, 1993.

[Paul]   H. Paul and J. Tindle, "Passive Optical Network Planning in Local Access Networks: An Optimisation Approach Utilising Genetic Algorithms," *BT Technology Journal*, Vol. 14, No. 2, April 1996.

[Piat]   G. Piatetsky-Shapiro and W. J. Frawley, *Knowledge Discovery in Databases*. AIAA Press/The MIT Press, Menlo Park, CA, 1991.

[Pohl]   J. Pohl, *Object Oriented Programming Using C++*. Benjamin Cummings, Menlo Park, CA, 1993.

[Pont]   C. Pontailler, "TMN and New Network Architectures," *IEEE Communications Magazine*, April 1993, pp. 84–88.

[Pryc]   M. de Prycker, *Asynchronous Transfer Mode: Solution for Broadband ISDN*. Ellis Horwood, Simon & Schuster, New York, 1991.

[Rao]   S. S. Rao, *Reliability-Based Design*, McGraw-Hill, New York, 1992.

[Rapp]   T. Rappaport, *Wireless Communications: Principles and Practice*. Prentice-Hall, Englewood Cliffs, NJ, and IEEE Press, Piscataway, NJ, 1995.

[Redl]   S. M. Redl, M. K. Weber, and M. W. Oliphan, *An Introduction to GSM*. Artech House, Boston, 1995.

[Robe 93]   T. G. Robertazzi, ed, *Performance Evaluation of High Speed Switching Fabrics and Networks: ATM, Broadband ISDN and MAN Technology*. IEEE Press, Piscataway, NJ, 1993, 450 pages.

[Robe 94]   T. G. Robertazzi, *Computer Networks and Systems: Queueing Theory and Performance Evaluation*. Springer-Verlag, New York, 1990, 1994.

[Rose]   M. T. Rose, *The Open Book: A Practical Perspective on OSI*. Prentice-Hall, Englewood Cliffs, NJ, 1990.

[Saad]   T. N. Saadawi, M. H. Ammar, and A. El Hakeem, *Fundamentals of Telecommunications Networks*. Wiley, New York, 1997.

[Sale]   J. Salemi, *PC Magazine Guide to Client/Server Databases*. Ziff-Davis Press, Emeryville, CA, 1993.

[Sall]   G. Sallai, "Planning Aspects of Network Protection Against Transmission Breakdowns," *Telecommunications Journal*, Vol. 53, July 1986, pp. 399–403.

[Sanc]   E. Sanchez and M. Tomassini, eds., *Towards Evolvable Hardware: The Evolutionary Engineering Approach*. Springer-Verlag, Berlin, 1996.

[Sasi]   R. Sasisekharan, V., Seshadri, and S. M. Weiss, "Data Mining and Forecasting in Large-Scale Telecommunication Networks," *IEEE Expert Intelligent Systems and Their Applications*, Vol. 11, No. 1, February 1996.

[Schw]   M. Schwartz, *Telecommunications Networks*, Addison-Wesley, Reading, MA, 1987.

[Sedg]   R. Sedgewick, *Algorithms*, Addison-Wesley, Reading MA, 1988.

[Semi]   "Seminar on Transmission Planning Aspects of Analogue–Digital Mixed Networks," *Telecommunications Journal*, Vol. 55, February 1988, pp. 101–106.

[Sevi]   G. Sevitsky, J. Martin, M. Zhou, A. Goodarzi, and H. Rabinowitz, "The NYNEX Network Exploratorium Visualization Tool: Visualizing Telephone Network Planning," *Proceedings of the SPIE*, Vol. 2656, January/February 1996, pp. 170–180.

[Shar 70]   R. L. Sharma and M. T. El Bardai, "Suboptimal Communications Network Synthesis,' *Proceedings of the International Conference on Communications*, Vol. 7, 1970, pp. 19–11 to 19–16.

[Shar 90]   R. L. Sharma, *Network Topology Optimization: The Art and Science of Network Design*. Van Nostrand Reinhold, New York, 1990.

[Sham]   N. M. Shamuyarira, "Telecommunications for Development of Rural and Remote Areas," *Telecommunications Journal*, Vol. 54, July 1987, pp. 455–458.

[Simp]   P. K. Simpson, ed., *Neural Networks Theory, Technology and Applications*. IEEE Press, Piscataway, NJ, 1995.

[Smal]   "R. B. Small, Debunking Data-Mining Myths," *Information Week*, January 20, 1997, pp. 55–60.

[Smit]   R. L. Smith, "Deferral Strategies for a Dynamic Communications Network," *Networks*, Vol. 9, 1979, pp. 61–87.

[Sore]   H. W. Sorenson, "Least Squares Estimation: From Gauss to Kalman," *IEEE Spectrum*, 1970, pp. 63–68.

[Spoh]   D. L. Spohn, *Data Network Design*. McGraw-Hill, New York, 1993.

[Stan]   J. Stanley, *Introduction to Neural Networks*, 2nd ed. California Scientific Software, Sierra Madre, CA, 1989.

[Stee]   R. Steele, *Mobile Radio Communications*. IEEE Press, Piscataway, NJ, and Pentech Press, London, 1992.

[Stei]   K. Steiglitz, P. Weiner, and D. J. Kleitman, "The Design of Minimum Cost Survivable Networks," *IEEE Transactions on Circuit Theory*, Vol CT-16, November 1969, pp. 455–460.

[Stew]   G. Stewart, *Introduction to Matrix Computations*. Academic Press, New York, 1973.

[Stro]   B. Stroustrup, *The C++ Programming Language*. Addison-Wesley, Reading, MA, 1991.

[Sysw]   G. Syswerda, "Uniform Crossover in Genetic Algorithms." In *Proceedings of the Third International Conference on Genetic Algorithms*, J. D. Schaffer, ed., Morgan Kaufman, San Francisco, 1989, pp. 2–9.

[Tane]   A. Tanenbaum, *Computer Networks*, 3rd ed, Prentice-Hall, Englewood Cliffs, NJ, 1996.

[Taub]   H. Taub, *Digital Circuits and Microprocessors*. McGraw-Hill, New York, 1982.

[Tind]   J. Tindle, S. J. Brewis, and H. M. Ryan, "Advanced Simulation and Optimisation of the Telecommunications Network," *BT Technology Journal*, Vol. 14, No. 2, April 1996, pp. 140–146.

[Toma]   M. Tomassini, "Evolutionary Algorithms." In *Towards Evolvable Hardware: The Evolutionary Engineering Approach*, E. Sanchez and M. Tomassini, eds. Springer-Verlag, Berlin, 1996, pp. 19–47.

[Tsui]   K. C. Tsui, B. Azvine, and M. Plumbley, "The Role of Neural and Evolutionary Computing in Intelligent Software Systems," *BT Technology Journal*, Vol. 14, No. 4, October 1996.

[Vans]   R. Van Slyke and H. Frank, "Network Reliability Analysis: Part I," *Networks*, Vol. 1, 1971, pp. 279–290.

[Whit 70]   V. K. M. Whitney, *A Study of Optimal file Assignment and Communication Network Configuration in Remote Access Computer Message Processing and Communication Systems*. Ph.D Thesis, University of Michigan, Ann Arbor, 1970, University Microfilms, Ann Arbor, 71-15345.

[Whit 72]   V. K. M. Whitney, "Lagrangian Optimization of Stochastic Communication System Models," *MRI Symposium on Computer Communication Networks*, Brooklyn, NY, April 1972.

[Wilk]   E. J. Wilkinson, "Planning the Use of Satellite Transmissions Within the South Pacific Telecommunications Development Programme," *Telecommunications Journal*, Vol. 54, November 1987, pp. 758–762.

[Xion]   L. Xiong-jian, "Network Planning Methodology and Practice in China," *IEEE Communications Magazine*, July 1993, pp. 34–37.

[Yage]   B. Yaged, "Minimum Cost routing for Static Network Models," *Networks*, Vol. 1, 1971, pp. 139–172.

[Zade]   N. Zadeh, "On Building Minimum Cost Communication Networks over Time," *Networks*, Vol. 4, 1974, pp. 19–34.

[Zupa]   J. Zupan, *Clustering of Large Data Sets*. Research Studies Press, a Division of Wiley, Chichester, 1982.

INDEX

ABOUT THE AUTHOR

Thomas G. Robertazzi received a Ph.D from Princeton University in 1981 and the B.E.E. from Cooper Union in 1977. He is presently an associate professor of electrical engineering at the State University of New York at Stony Brook. During 1982–1983, he was an assistant professor in the electrical engineering department of Manhattan College, Riverdale, New York. He was a visiting research scientist at Columbia University's electrical engineering department during Fall 1990.

In recent years, Professor Robertazzi has taught engineering courses at Stony Brook, Cooper Union, and in industry. Among topics covered in these courses are network management and planning, networking, performance evaluation, wireless technology, and communications systems. Since 1993 Professor Robertazzi has also been faculty director of the Stony Brook Interdisciplinary Program in Science and Engineering. This program is based in a residential undergraduate college, and provides an academically enriched environment for the college's residents.

Professor Robertazzi's research interests involve the performance evaluation of computer and communication systems. He has published extensively in the areas of parallel processor scheduling, ATM switching, queueing networks, Petri networks, and multihop radio networks. Along with Dr. James Cheng, Professor Robertazzi is the cocreator of divisible load models of parallel processor scheduling.

Professor Robertazzi served as editor for books for the IEEE Communications Society and as an associate editor of the journal, *Wireless Networks*. He has authored one book, coauthored a second, and edited a third in the area of performance evaluation.

R JOHNSON
PSC 78 BOX 3369
APO AP 96326